The Exploding Universe

Nigel Henbest

Macmillan Publishing Co., Inc.
New York

Macmillan Publishing Co., Inc.
866 Third Avenue, New York, N.Y. 10022

Library of Congress Cataloging in Publication Data
Henbest, Nigel.
    The exploding universe.
    1. Astronomy—Popular works. I. Title.
QB44.2.H44 1979   520   79-10103

ISBN: 0-02-550920-9

First American Edition 1979
Printed in Hong Kong

Introduction

'The proper study of mankind is Man,' wrote Alexander Pope in the eighteenth century. But by definition mankind is an inquisitive species: we have never stopped studying our environment. We cannot understand our lives without knowing how and why life began, or at least asking the question. The question leads to the Sun, which leads to the entire Universe. Now at last, in our century, the answers are beginning to come.

It can also be said that mankind is an instinctively scientific species. From earliest times we have tried to list and classify the things we see, as though knowing that this is the only way to understand them: hence the ideas of the first scientists have survived to be confirmed. Ancient philosophy knew no limits; its speculations ran freely into areas we would now describe as scientific. Epicurus (341–271 BC) acquired a reputation for atheism and hedonism because he was a materialist, although self-control and moderation were central to his philosophy. He also formulated one of the first atomic theories. Empedocles (fifth century BC) taught a theory of four fundamental elements; he might not be displeased by our theory of four fundamental forces.

Kepler gave the first substantially correct description of the solar system in 1609; he formulated laws of planetary motion which Newton later showed to be the result of gravitation. In his *Harmony of the World* (1619) he examined and compared mathematical relationships, such as in planetary orbits, musical intervals and geometry. We say that the driving force behind his quest for discovery was religious, yet he was persecuted by religious authorities in his lifetime. The truth is that a great mind always has religious ideas that transcend its relatively superstitious contemporaries.

So it is in modern times: all knowledge must be related. Physicists used to believe in something called ether; believing that nothing, not even light, could travel through a vacuum, they postulated that space must be filled with a mysterious substance which had the contradictory properties of being far more rigid than steel but impossible to detect. It was an experiment designed to prove or disprove the existence of this substance which led to Einstein's Theory of Relativity. The idea of ether had to be discarded, yet many scientists could not give it up until their dying day. Similarly, Max Planck, whose Quantum Theory has done as much to alter our understanding of physics as Relativity, spent the rest of his life trying to reconcile his own discoveries with the classical physics he loved. It is clear that any theory may have to be modified or discarded in the light of new evidence.

The path of knowledge described by the great minds of history is being pursued today, by Cambridge University's Stephen Hawking and many others around the world, aided by techniques and equipment of astonishing sophistication. The philosophical and mathematical elegance of their discoveries will be our most important bequest to our descendents. Yet physicists also say that there is a natural limit to what we can know: their theories must allow for Uncertainty, as well as events which demand the suspension of disbelief. The true man of science learns humility in the face of Creation.

This would not have surprised the theologian Teilhard de Chardin. 'Everything that rises must converge,' he wrote, believing that the paths of knowledge do not run forever parallel, but lead upward and meet. This book is a summary of today's knowledge of Creation, and it is also a tribute to the majestic beauty of the Universe, and to the beauty of Truth, which is perhaps the same thing.

DC

Contents

Sunrise over the seashore evokes the four 'elements' of the ancient philosophers: earth, water, air and fire. Modern discoveries have unravelled the links between these different kinds of matter and have shown us how the Sun 'burns' so fiercely; and they have taught us that our Earth is only one small planet in an immense Universe.

1: Exploring the Universe

Lazing on a warm beach on a summer's day, it's not easy to concentrate. Thoughts flow through our minds and out again, broken by the shouts of children, lulled by the swishing throb of the sea. On the beach, everyone is a philosopher. We vaguely wonder why the Sun should be so hot, while the sea seems so cold – and, as often as not, the thought drifts away before we try to answer it. But let's catch some of these questions before they fade.

Why is the Sun hot but the sea cool? Why is water liquid but sand solid? What are they made of – and what is air, this invisible substance that we must breathe to keep alive? What is the difference between life and non-life? Are there other living creatures far out across the Universe? Is there an edge to the Universe? How long ago did it start – and how?

We've all had thoughts like these, and the answers can usually be found in an encyclopedia or a reference book. This book isn't an encyclopedia, nor is it primarily a reference book. It is an exploration. We shall explore the make-up of the Universe, from sand grain to lightning stroke, from atom to quasar, asking 'Why is it the way that it is?' Man's present knowledge is surprisingly complete, even when it comes to explaining the very large in terms of the very small, the very powerful in terms of the unimaginably weak. On the way, we shall throw some light on how Man has satisfied his questing instinct, and wrested Nature's secrets from her.

The art of science

Generations of men and women have spent entire lifetimes pursuing the solutions to seemingly unanswerable questions about the Universe – such questions are the essence of science. To understand the Universe, and all that's in it, is one of the greatest aims of Man. The pursuit of science (which literally means 'knowing') is as much an art in its own way as the composition of a great symphony, and is full of endless fascination for the seeker after knowledge. There are those who feel that the scientific discoveries of the twentieth century will be seen, by future generations, as the supreme esthetic achievement of our age.

Despite this, 'science' has often had a bad press recently: scientists are popularly seen as cold and unemotional, while scientific results appear to threaten our very existence. Such attitudes are partly a backlash from the euphoria of the 1950s and '60s, when scientific discoveries seemed poised to solve all the world's problems. The failures here, however, have been mainly in the realms of economics and politics: often it doesn't pay – financially or politically – to use the science and technology which is ready and waiting. On the other hand, one scientific discovery which *has* been taken up avidly by the 'developed' nations, in spite of its huge expense, is the terrifyingly destructive power of atomic energy in the H-bomb. It is sad that the name of science is often blackened by association with that age-old ogre, the misuse of knowledge.

It would certainly be too glib to say that scientists merely make discoveries, and the rest of us decide whether they will be used for good or evil. There's a simmering controversy in science at present over 'genetic engineering' – recombinant DNA research, to give it its proper title. In future this may allow us to doctor human cells so that we can choose the sex of our children, but government laboratories may also be able to turn out thousands of identical supermen to serve as soldiers. Even at present there is a danger that the escape of just *one* dangerously mutated cell, microscopically tiny in itself, could start an irreversible epidemic which would endanger all human life on this planet. Scientists in genetic engineering laboratories have an added responsibility to all mankind.

Yet Man is consumed by the passion to find out. The scientist will always be with us, pursuing knowledge in his own way, and as soon as we begin to ask about the real world, we are following the pursuit of science ourselves.

More and more complex

Unfortunately, the pace of scientific research has taken the professional scientist further and further from the rest of us. A century or two ago, an educated man could read and understand first-hand accounts of the latest research in the journals; revolutionary scientific results were indeed often published as books aimed at the general reader – for example, Charles Darwin's *Origin of Species*. Today, the journals are so specialized that only someone with years of experience can hope to understand most research papers, and then only in his own field. A trained astronomer has nearly as much difficulty as anyone else in tackling a biology journal.

It's important that the growing gap between the working scientist and the public should be bridged. Many of the products of advanced science and technology reach us: non-stick frying pans, pocket calculators and digital watches, to mention a few. But what scientists actually do, and how they do it, is a mystery to most of us. One problem is simply that too much is known. The English genius, Sir Isaac Newton, once said of his remarkable achievements that if he had seen further than other men, it was because he stood on the shoulders of giants. Learning the total knowledge of previous generations, the past giants of science, and finding out the best way to pursue knowledge, are processes which take years to complete. The old-fashioned school approach, that science consists of remembering facts, is a necessary step on the way to the frontiers of knowledge; if the school experience is often a matter of staving off boredom, perhaps the scientific method is not entirely to blame.

Unmanned spacecraft are exploring regions of the solar system which man has yet to reach. This Viking spacecraft, launched from Cape Canaveral in 1975, carried an automatic laboratory to search for life on Mars.

The moment of discovery

But for scientists at the forefront of research, the whole complexion of science takes on a richer hue. With the background knowledge safely tucked away in one's mind, the assault on Nature's mysteries becomes a genuinely creative event. Whatever the research field, there are always unexplored incomplete areas, containing loose ends to be tied up. The ultimate moment for any scientist is when these inexplicable facts are seen as part of a whole: like the musical resolution of discords, they are found to be steps towards a new harmonious vision. The light can dawn at the most unexpected times: for Archimedes ('Eureka' – 'I have found it!') in the bath; and for the chemist Kekulé, who saw in a dream how benzene molecules are constructed, while dozing on a bus. In retrospect, the startling new theory always seems obvious, like a crossword solution, but it is only a small part of the story. An inspiration only comes after months or years of work, and it must be followed by more experiments to test the new ideas. The result is a piece of knowledge, a corner of the Universe illuminated.

A fifteenth century alchemist and his assistants at work. Modern chemical laboratories contain some apparatus developed from the alchemist's, like the still on the lower left.

At its highest level, science is an incredibly exciting human activity, and more than rewards the years of study leading up to it. The real breakthroughs are made only by scientists with tremendous imagination, who can see through the complexities of an experimental result, or through the tangled web of previous theories, and select only the most important points to build on. 'Cold' and 'unemotional' are hardly the words to describe a top-flight scientist – he is a passionate seeker after knowledge, an inspired dreamer.

But after the excitement of a new discovery, a scientific breakthrough, it's not always easy to explain its importance to the world at large. Indeed, it is true to say that a great many scientists are far happier getting on with their research than in explaining their advances to the general public – even when it is the latter who is paying for the research. Astronomy usually fares better than the other sciences (apart from biology) because astronomical discoveries are comparatively easy to understand. It wasn't too difficult to see that the pulsars – celestial objects emitting a regular 'ticking' radio signal – were tiny, rapidly spinning stars; and the discovery made all the newspapers. But when physicists found a fourth quark – one of the basic building blocks of matter – it didn't make much of a ripple in the popular press. A certain amount of astronomy is part of everyone's general knowledge, but popular science books and school syllabuses haven't yet caught up with the explosion of knowledge in the subatomic world in recent years.

From pulsars to particles

There is certainly as much excitement in the world of the unimaginably tiny as there is in the astronomers'. Physicists believe they may at last be able to say what the most fundamental particles – the smallest, most basic building blocks of the Universe – are, and why they behave as they do. They can show how these particles (including the quarks just referred to) build up into the atoms of everyday matter, while the forces between particles give rise to the power of electricity, of the H-bomb, and of the gravity holding us down to the Earth's surface.

Perhaps the most surprising fact of all is that matter is the same, and behaves in the same way, throughout the entire Universe. Astronomers' discoveries can all be explained by exactly the same kinds of particles, the same forces, and the same laws of Nature that physicists find in a laboratory on Earth. That seems all the more remarkable when we meet some of the bizarre objects which astronomers are now coming across: whole stars condensed into a ball only a few kilometres across; black holes, where gravity can pull matter out of sight of the rest of the Universe; the raging holocaust in the heart of a quasar; and radio waves lingering on from the birth of the Universe itself.

Although this book is about the Universe, it is not simply the Universe as seen through astronomers' eyes: ours will be an exploration with a difference. Throughout, we shall see that everything in the Universe is made from a few fundamental particles, influencing each other by a few forces: everyday objects like chalk and cheese, beach balls and television sets; the rather more complicated objects we call 'alive'; and then the rest of the Universe, outside our own parochial, though seemingly important, little niche. In asking 'Why is the Universe the way it is?', we shall find that present-day science has a remarkably consistent set of solutions, but also that there are still many questions for the scientist to answer. Part of our quest must be concerned with looking down some of the paths still to be explored, and wondering how they may change our ideas about the Universe.

Elements: the basic substances

The ancient Greeks were the first people to try to find solutions to the riddles of Nature, but they lacked precise instruments with which to test their theories, and their ideas generally stood or fell by their philosophical acceptability rather than by experimental confirmation. The front runner

among their theories of matter was that put forward by the philosopher and mystic Empedocles in the fifth century BC. He taught that the Universe is composed of four elements – earth, water, air and fire – and that all the various kinds of matter we know differ simply in the proportions of the elements making them up. On Earth we never come across these elements in a pure state: for example, ordinary water would be mainly, but not entirely, composed of the element 'water'. As well as his elements, Empedocles' world contained two opposing forces, love and hate, which governed inanimate objects as well as living creatures.

There is certainly some logic behind his scheme, particularly in his notion that the elements naturally occur in a particular order: earth is lowest, above it is the natural place of water, higher still is the realm of air, and above all the others comes fire. Since each element seeks its own place, a clod of earth falls down through air and water, and drops of water fall through air; while on the other hand, air bubbles rise through water, and fire shoots upwards in air. The gods (if they existed) must live in the fiery uppermost layer – and there was a popular story that the recluse Empedocles, after retiring to a solitary life on the upper wastes of Mount Etna,

eventually believed that he himself was a god, and threw himself into Etna's fiery crater.

Empedocles' ideas were popular for two thousand years; indeed, we still use them figuratively today, when we speak of a violent storm as the 'raging of the elements'. The great Greek philosopher Aristotle incorporated the four elements into his comprehensive scheme of the Universe, and included a fifth, more sublime element: the 'ether'. This lies above the layer of fire, and does not come down below the orbit of the Moon, for it is incompatible with the corrupt Earth.

Throughout the Middle Ages in Europe, Aristotle was the guiding light for men of learning. His theories were reconciled with the Bible by St Thomas Aquinas, and thereafter the mediaeval scholars' method of answering scientific questions was simply to consult the works of Aristotle. Only the exceptional scholar thought of actually performing an experiment. For example, Aristotle stated that a heavy rock falls faster than a light one, and it took a genius of the stature of Galileo actually to drop two rocks – and prove Aristotle wrong. (By the way, it is unlikely that he dropped them from the picturesque Leaning Tower of Pisa, as the popular myth insists.)

Alchemists were meanwhile having a field day with the four accessible elements. They had a particular interest in the structure of matter, for the goal of alchemy very quickly centred on the transformation of 'base' metals into the most noble metal, gold. If different kinds of matter were just differing mixtures of the basic four elements, then transmutation would simply involve altering the proportions. The approach to the 'Great Work' was outlined by the Arab, Jabir ibn Hayyan, known in western Europe as Geber. Sulphur – choking, yellow brimstone – was the essence of combustibility (fire), while the liquid metal mercury, or quicksilver, was the essence of fusibility ('wateriness'). Combine impure forms of these, and base metals, like copper or lead, would result; purify them before combining, and gold would be produced. At an even higher level of purity, the combination would end up as the Philosopher's Stone. This 'stone', the ultimate aim of the mediaeval alchemists, would by itself transmute base metals into gold; and, by analogy with this 'purification' of metals, it could purify the human body and soul. By eliminating all illness, including old age, the Philosopher's Stone promised immortality, and it gained the alternative name of the Elixir of Life.

The attraction of the alchemists' ideas on human minds is immense. The fabric of alchemy only began to crumble three centuries ago, at a time when the laws of physics and the basic ideas of astronomy were already established by Sir Isaac Newton and his predecessors. Serious alchemists continued to practise until about 1800, and it is said that Adolf

An immense source of energy resides in the central nuclei of atoms, as shown by this United States atomic bomb test at

Bikini Atoll on 1 July 1946. Note the captured warships (lower left) exposed to the effects of the blast.

Hitler employed an alchemist in the hope of bolstering the gold reserves of the Third Reich.

The most astute minds of the seventeenth century, however, had realized that everyday substances cannot be broken down into the earth, water, air and fire that they were supposed to be made of. The main reason for believing in Empedocles' elements was the immense awe with which mediaeval man regarded the ancient philosophers. In the 1660s, the Anglo-Irish scientist Robert Boyle – perhaps the first man who was more 'chemist' than 'alchemist' – suggested that the 'elements' of which all other substances are made are ordinary substances themselves. On this modern view, the liquid metal mercury and the yellow non-metal sulphur, for example, are both elements, for they cannot be broken down by chemical reaction into anything simpler. They can react together, however, to form the red compound cinnabar (mercuric sulphide) – but not to make gold, which is an element in its own right.

The list of chemical elements was extended in the early nineteenth century by Sir Humphry Davy, the inventor of the miner's safety lamp. By passing an electric current through hot, molten compounds, he decomposed them into their elements. Ordinary kitchen salt broke down into the poisonous gas chlorine, and a new metal, sodium, so reactive that it will fizz and dissolve instantly when thrown into water. To check this analysis, put these two fierce elements together, and in a flash you are left with sodium chloride: cooking salt.

In Davy's time, 47 elements were known; the total now stands at 105. Only 88 are found in Nature, the others being man-made in nuclear reactors. Many of the latter can exist for only a short time, and their existence is proved by the physicist rather than the chemist.

Atoms: the basis of elements

At the same time as Davy was breaking down our everyday substances to their basic elements, it was becoming clear that these elements were not made up of continuous ('solid') material, but were composed of huge numbers of tiny individual particles, called atoms. The atomic theory of matter had been around since Greek times; but it had not found favour with Aristotle, and had consequently been ignored in mediaeval Europe. Chemists at the turn of the nineteenth century were finding that elements combine in certain definite proportions. On the old view that substances were continuous – you could cut a sand grain in half an infinite number of times, and each successively tinier piece would always be sand – the chemical elements should be able to combine in any proportion. The new, exact chemistry experiments, pioneered in the eighteenth century by the father of modern chemistry, Antoine Lavoisier, showed that there must be a smallest particle of each element, and that these atoms must combine in particular, simple ratios to make up the chemical compounds we meet in everyday life, such as salt, sand or water. Thus the architecture of matter was elucidated. If we could cut our grain of sand small enough, we should find its basic unit to consist of two atoms of the element oxygen combined with one atom of another element, silicon. This simple trio is the basic molecule of silicon dioxide, the compound of which ordinary sand consists, and it is repeated countless times – millions of millions of millions – to make one grain of sand among the multitude on the beach.

Beyond atoms: particles

But the power of electricity, which had enabled Davy to tear apart the sodium and chlorine atoms in salt, was next to be associated with the stripping off of the outer layers of the atoms themselves. Electrical discharges in a very tenuous gas produced a shower of 'cathode rays', streams of subatomic particles with a negative electric charge. These *electrons* (as we call them today) are the minute projectiles which are fired from the back of the tube in a television set, to produce the light and dark pattern as they are scanned across the phosphor screen. When they were discovered in 1897, they were an enigma – where did these electrically charged particles fit into the scheme of chemical elements?

Only after the dust had settled from the scientific revolution of the early twentieth century, which produced the first fundamentally new theories of the physical world since Newton's time (quantum theory and the theory of relativity), did the role of electrons become clear. Atoms are not just miniature billiard balls, hooking together to make up the molecules which comprise everyday matter. They have an internal structure: the negatively charged electrons in the outer part of the atom are orbiting around a central, positively charged nucleus, rather like planets spinning around the Sun (though not so unchangingly: for this reason an atom's collection of electrons is often referred to as a 'cloud'). The nucleus itself is composed of two types of particle, positively charged protons and electrically neutral – uncharged – neutrons. The Universe seemed to be remarkably simple in its construction, requiring only three different fundamental particles – electrons, protons and neutrons – which come together to form atoms. The different chemical elements are distinguished by having different numbers of protons in the nucleus, and a correspondingly different number of the oppositely charged electrons in orbit – the number of protons and electrons in a single atom must be equal, so that their electric charges cancel out. Since it is the behaviour of the outermost electrons which controls how atoms join together in molecules, the different elements will differ in their chemical properties.

A simple world, indeed; and the structure of the Universe seemed solved by the 1930s. But experimental physicists continued probing the sub-atomic world, and soon their results began to shake the theorists' complacency. Although there was no doubt that the structure of the atom was in the main correctly understood, it was soon evident that there were many other 'fundamental' particles. Most of these new particles were not stable for more than a fraction of a second, 'decaying' rapidly into other different particles, and giving off radiation. They were clearly different from the electron, proton and neutron. To the embarrassment of particle physicists – as those studying the subatomic world became known – the number of particles grew steadily through the 1950s. Whereas it once seemed a remarkable achievement to be able to explain the hundred-odd elements in terms of just three fundamental particles, a terrible irony became apparent as physicists realized that the proton, neutron and electron are only three of over a hundred different particles which can exist in Nature.

Particles of particles

Can all these 'fundamental' particles be made up of just a few even tinier particles? If so, can we explain why only a few are stable, and most of Nature's subatomic particles quickly break down into others? Since 1960, particle physicists have been obsessed with these questions, and the basic scheme of Nature has now, it appears, again turned out to be relatively simple. The proton and neutron, and the vast majority of the other subatomic particles, are held to be made up of combinations of just four different *quarks*.

The leading American physicist, Murray Gell-Mann, chose the term 'quark' in 1964, when it was believed that only three kinds of basic particles were needed to make up the heavier particles ('Three quarks for Muster Mark,' calls James Joyce in *Finnegans Wake*). Gell-Mann's colleague, George Zweig, proposed to call them 'aces', but the name never caught on – even though a fourth quark has now been 'found'. (Indeed, latest experiments indicate that another one may be around, and possibly a sixth.)

The four quarks make up the vast majority of subatomic particles, when combined two or three at a time. There are four particles which stand apart, however. The electron, its heavier brother the muon, and two types of neutrino constitute a separate family, the leptons. Unlike quarks, leptons do not band together to make heavier particles, but live separate and solitary existences.

To answer the question 'What is everything made of?', the modern physicist would reply thus: the Universe consists ultimately of only eight kinds of particle, which fall into two camps: four quarks, banded together to make heavier particles like protons and neutrons; and four leptons, including our old friend the electron. Combine protons and neutrons together in a nucleus, and clothe it in orbiting electrons to form an atom; put atoms together in molecules, joined up by their swirling electron clouds, to build up everything in our everyday world.

The four forces

And what of Empedocles' forces – love and hate? They have fared slightly better than his elements; for forces are the second ingredient of the Universe as scientists see it today. Every fundamental particle in the Universe is affecting the motion of every other particle – attracting it or repelling it – by one or more of the four forces known to present-day physics. Two of these forces are familiar to all of us: gravitation and electromagnetism; while the other two, nuclear, forces operate only at the scale of atoms themselves.

Let go of the book and it will fall to the floor: we say Earth's gravity has pulled it down. In fact, every particle in the book is attracting every particle in the Earth, and vice versa, and it is their mutual gravitational tug which pulls them together. Naturally the book moves much more than the Earth, for it contains so much less matter; but the Earth moves too, rising one ten-thousandth of a millionth of a millionth of a millionth of a millimetre to meet the falling book.

Gravity holds us tightly to the Earth: to escape from Earth's hold, spacecraft must reach the tremendous speed of 11 kilometres (6·82 miles) per second – 40,000 kilometres (24,855 miles) per hour. Our globe itself is kept in orbit around the life-giving Sun by the latter's powerful gravitational attraction; and on a larger scale still, the force of gravitation constrains our Sun to a circular orbit around our Galaxy of 100 thousand million stars (all held together by their mutual gravitational pull), despite its speed around the Galaxy's centre of some 250 kilometres (155 miles) per second.

So it's rather surprising to learn that gravitation is by far the weakest of Nature's forces. On an astronomical scale, it is supreme only because it is a long-range force, and attracts everything. The two nuclear forces, though much more powerful, operate only between particles that are extremely close – as near to one another as the protons and neutrons in the nucleus of an atom. The electromagnetic force is also far more powerful than gravity, and it does have a long range. In practice, however, the forces from the positively charged protons and from the negative electrons cancel out, once we are dealing with a piece of matter much larger than an atom.

The tremendous strength of the electromagnetic force – 'electromagnetic' describes both electric and magnetic forces, which are in fact closely related – can be shown by a simple experiment. Rub a plastic comb on your sleeve and hold it just above a few tiny scraps of torn-up paper: you will see the tiny pieces jump upwards. In the act of rubbing, a minute proportion of the electrons in the atoms making up the sleeve has been transferred to those in the comb, and this has given it an overall negative electric charge. Even though the extra electrons amount to only one for every million million atoms in the comb, this tiny concentration of electricity creates enough force to pull the paper scraps upwards, against the gravitational pull of all the atoms in the Earth.

The first fact we learn about electric forces is that 'like charges repel' and 'unlike charges attract'. Thus a positively charged proton attracts negatively charged electrons, but two protons will repel one another. It is the attraction between protons and electrons which keeps the electrons in atoms orbiting around the nucleus, just as on a much bigger scale, the always attractive force of gravity keeps the Earth circling the Sun, and the Moon orbiting the Earth. To compare the relative strengths of gravitational and electric forces, suppose we could completely destroy electrons at will. If we then reduced a single sand dune on the Earth, and one on the Moon, to their constituent protons, the repulsion between the positively charged dunes would exceed the gravitational attraction between the Earth and Moon, and would force our Moon off into space.

Returning to more mundane affairs, it is the electric force which causes electrons to flow as an electric current along a wire. This property is vital to us, for without electricity twentieth-century life as we know it would be impossible. In a metal, electrons are not tightly bound to their own atoms, and a slight electric force – a voltage – causes them to drift along a metal wire. There are always as many electrons as protons in the wire, so this drift does not make the wire itself positive or negative. The flow of electrons carries energy, however, rather as a stream of water does, and this energy can be used for the wide variety of purposes that we associate with the word electricity: heating, lighting and broadcasting, to name but a few.

A moving electric charge also produces the rather mysterious force of magnetism. Historically, this force was first found to be associated with *lodestone*, a natural mineral – what is now called magnetite, a form of iron ore. This was given its name (which means 'course stone') because it pointed roughly north–south when freely suspended. Compasses made with lodestone, and later with metal magnets, were invaluable to European sailors from the thirteenth century onwards. Magnetism has always seemed rather like electricity, but with subtle differences. Like electric charges, there are two types of magnetic *pole*, named 'north' (N) and 'south' (S) after the direction they point in when suspended as a compass; and again, like poles repel while unlike poles attract. But magnetism differs in that poles can never exist in isolation. A magnet never has just a N pole, for example, but always has N and S poles of equal strength. Cut a magnet in half, and the cut ends become poles opposite to the existing pole on each piece.

When electric currents were first extensively studied, at the beginning of the nineteenth century, a curious link was found between electricity and magnetism. The moving electric charges in the current create a magnetic field around the wire, and conversely a current could be started simply by moving a conducting wire through a magnetic field. The first effect is used to great practical advantage in the *electromagnet*, where a coil of wire carrying a current behaves exactly like a huge magnet. With an electromagnet, however, the operator can turn off the magnetic force by simply switching off the electric current – invaluable when shifting around old cars in a scrapyard, for example.

The second effect is even more important to our everyday lives, for it is the principle of the electric generator. In a power station, energy from burning coal or oil, falling water or a nuclear reactor is used to revolve a coil of wire in a magnetic field. The mechanical energy used to turn the generator (or dynamo) is converted into electric current by the spinning coil. In this form it can be sent out via the electricity grid, and reconverted to mechanical energy – or light, heat, etc. – in the home or factory.

The behaviour of permanent magnets became clear when it was realized that an atom consists of electrons whirling around a central nucleus. The electrons constitute a circulating current, like the flow of electrons around the coil of an electromagnet, and so each atom is itself a minuscule electromagnet with N and S poles. In fact, the circulation of the multitude of electrons in most atoms tends to produce magnetic fields which tend to cancel out; but in a handful of elements – of which iron is the most common – the individual atomic electromagnets are not only strong, but can be lined up throughout a solid block of the substance, N pole to S pole all the way through, with the effect that one end of the block acts as all S, and the other as all N.

Electricity and magnetism are thus different manifestations of the same fundamental force, electromagnetism. This, the second most powerful of the four forces in Nature, binds electrons to atoms, and builds up molecules of everyday substances by the interactions of the electron clouds with those of neighbouring atoms. Our world of everyday objects is very much an electromagnetic world.

The two remaining forces – the 'strong' and the 'weak' nuclear forces – may seem at first to be remote from everyday life; yet they are as important as gravity and electromagnetism in determining the structure of the Universe.

The strong force holds together the quarks. It keeps them bound up as twos or threes in the old fundamental particles, and with such a strength that physicists have not yet succeeded in smashing these particles apart into naked, isolated quarks. Almost as an afterthought, the remnants of this colossal force inside the proton and neutron leak out to hold these particles together in the nucleus of an atom.

The need for some kind of attractive force in the nucleus was obvious as soon as physicists found that many protons could live together in the same nucleus. When these positively charged particles are held so close together, the electric repulsion between them is immense – two protons in the tiny nucleus of an iron atom repel each other with a force equal to the weight of this book. Obviously some far stronger force is at work in the nucleus, binding the protons (and neutrons) together, and it must be some hundred times stronger than electromagnetism.

Some aspects of the strong nuclear force are pretty familiar to us. The powerful attraction of protons and neutrons can be released in deadly fury in the hydrogen bomb, as new nuclei are built up from smaller ones. The same process – the nuclear *fusion* of hydrogen – occurs in the core of the Sun and other stars, although here the strong force, its energy converted to heat and light, comes to us from a safe distance as a benign influence.

The weak force has for long been the most mysterious of the four forces of Nature. It has no obvious effect on the everyday world, and even in the modern physics laboratory its actions are usually masked by the thousand-times-stronger electromagnetic force, or by the even more powerful strong force. Most of the effects of the weak force are manifest in the reactions it causes, rather than in actual attraction or repulsion of particles. A free neutron, for example, only lives for 15 minutes on average. Then it breaks down into a proton, an electron and an elusive antineutrino (the antimatter equivalent of the 'ordinary' neutrino), and this reaction is mediated by the weak nuclear force – theory shows that it could not happen if gravitation, electromagnetism and the strong nuclear force were the only forces in Nature. In practical terms, the weak force is perhaps most important in governing the first stages of hydrogen fusion in stars: without it, stars would never

The forces of nature are all related. Electricity and magnetism are closely linked, and this electromagnetic force has now been related to the weak nuclear

force. Physicists are confident that the links with the strong nuclear force and gravitation will eventually be found.

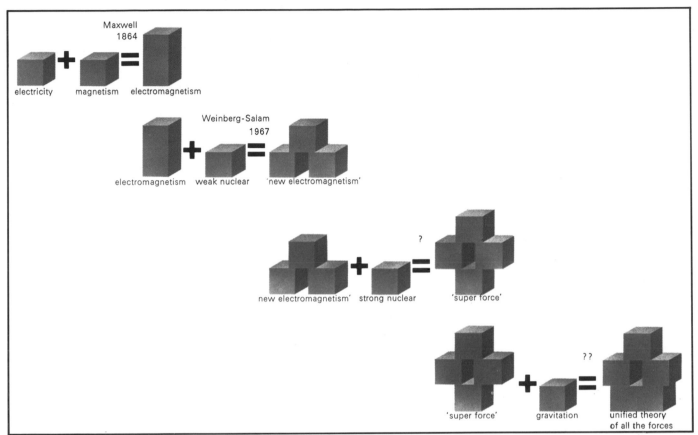

shine, or might explode as soon as they were born, as we shall see when we take a closer look into the centres of the stars in Chapter 8.

The weak force, however, may be just another aspect of the electromagnetic force. We have seen how the apparently mysterious force of magnetism turned out to be another aspect of the electrical force; and in 1967 Steven Weinberg and Abdus Salam proposed that the weak force may be related too. The theory underlying this unification is not simple; but it does seem as though Nature may have only three forces, one of which – the 'new electromagnetism' – can show itself in several ways: electric, magnetic, or weak nuclear forces. Indeed there is hope among particle physicists that in only a few years, the strong force may also be incorporated into this scheme as just another manifestation of a single force. This 'superforce' would combine strong, weak and electromagnetic forces as different ways of describing its all-embracing effects. Only gravitation still presents an outstanding problem in the reconciliation of forces.

The laws of Nature

Eight particles, and four forces (perhaps soon to be reduced to two?) are the basic stuff of the Universe, as seen through modern scientists' eyes. There is one additional, more subtle, ingredient: the laws of Nature. If we wanted to rid

ourselves of the Moon by the electrical repulsion of two dunes, we would have to destroy a vast amount of negative electric charge in the form of electrons. All experiments which have ever been performed show that Nature does not allow us to destroy electric charge. Negative charge can be neutralized by adding an equal positive charge, but electric charges cannot be created or destroyed. This 'law of conservation of electric charge' is just one example of the laws of Nature, which govern how particles behave.

The term 'laws' of Nature is something of a misnomer, for they are really just generalizations about how the Universe behaves. It takes thousands of experiments to determine a particular 'law'; while an established 'law' would have to be given up or modified if even a single experiment were to break it. Unlike the case of human laws, when scientists find a reaction which transgresses against an established scientific 'law', it is the 'law' which faces the penalty, not the culprit! Laws of Nature come in many kinds, often of the 'conservation' type which applies to electric charges. They show that the natural world has order behind it. The particles and forces make up the skeleton and muscles of the Universe, but they would just form chaos unless there were some architecture to fit them together.

And now, let us begin our exploration, for the whole vastness of the Universe is waiting. . . .

A large gas cloud (nebula) in the constellation Serpens is lit up by the very bright, newly formed stars visible near the centre. Dust clouds are silhouetted as dark globules and 'elephant trunks', they may condense as planets around the forming stars. Copyright by the California Institute of Technology and the Carnegie Institution of Washington. Reproduced by permission from the Hale Observatories.

2: Surveying the Depths of Space

Our planet Earth is a mere speck in the boundless Universe. From out in space, the dim disc of Earth is overwhelmed by the radiance of the Sun, our local star, provider of all our light and warmth. Without the Sun, Earth's surface would freeze to −270°C (−454°F), near the absolute zero of temperature. Yet the Sun is fairly inconspicuous in the league of stars. Glance out on a clear night, and you will see stars ten thousand times more powerful – but reduced to gently twinkling points of light because of their vast distances from us.

The patterns of stars in the night sky make up the constellation patterns of the ancients, named after heroes (Orion; Hercules) or animals (Taurus, the Bull; Delphinus, the Dolphin). In the confusing multitude of stars visible on a clear night – some three thousand at any one time – the constellations still represent the easiest way to learn one's way around the sky, and to pick out individual stars. Although we now know that all the stars in the 'join-the-dots' pattern of a constellation are generally unrelated – some will be comparatively nearby, faint stars, while others are distant, brightly shining beacons – they are still used by astronomers to define approximate regions of the sky. Astronomy, though a thoroughly modern science, still contains intriguing references to its past. As technology has advanced, and has enabled astronomers to pick up other radiations coming from the heavens, the contrast has heightened. Sensitive instruments sent up on a rocket flight in 1962 discovered X-rays coming from a point in the sky. The position could not be measured accurately then; but the general direction was from the constellation Scorpius, and the source was named Scorpius X-1, the first X-ray source in the constellation of the Scorpion.

To the modern astronomer, the stars are more than dots of light, or just fixed points in the sky, convenient for navigation. The battery of instruments available to today's astronomers has made the other stars almost as familiar as the Sun. And the huge range in stars is staggering. Although some outshine the Sun thousands of times, many more are far fainter than our star. Some are hundreds of times larger than the Sun, but their material is spread out through their huge volumes so tenuously that most of the star is a million times thinner than the air we breathe. Other stars are minute, with as much matter as the Sun compressed into a body the size of the Earth.

Families of stars

Many stars come in pairs, endlessly orbiting one another. In some close binary (double-star) systems, one star's gravitational attraction tears off the outer layers of the other, pulling them down to its surface, or spinning streamers of hot gas out into space. Other stars are still more gregarious, and live in huge clusters of up to a million stars, all orbiting around one another like a cosmic swarm of bees.

Each star, including the Sun, belongs to a single system of stars, a *galaxy*. Our own galaxy is a huge, flattened disc of stars, rotating around a central 'hub' which lies in the direction of the southern constellation Sagittarius. Looking around the sky, we see many more distant stars when our line of sight lies entirely within the disc of our galaxy than when we are looking out in directions away from the disc. The disc of our galaxy therefore appears as a faintly shining band of light around the sky, the summed-up light of millions of distant stars too faint to be seen individually. The perspective effect of fainter (more distant) stars crowding more closely to this glowing band can be seen on a clear night, especially in southern latitudes where the central 'hub' is visible. The ancient Greeks ascribed the faint white band to milk spilt from Juno's breast, and their name 'the Milky Way' (*gala* is Greek for 'milk') has stuck ever since. Our galaxy, of which the Sun is just one insignificant member, is often referred to as the Milky Way Galaxy, or simply, the Galaxy.

Other galaxies of stars exist far out beyond our own. Some are flat, like the Milky Way Galaxy; others are round; and some are completely irregular in shape. All of them are composed of stars which are basically no different from the stars of the Milky Way. The study of other galaxies has taught us that our Galaxy is a very typical specimen, just as the Sun is a very typical star within the Galaxy, and the Earth is a quite ordinary planet. Modern astronomy has certainly taken us far from the mediaeval viewpoint that the Earth is the centre, and most important part, of the Universe.

Ironically enough, it is much more difficult to tell how the stars are arranged in our Galaxy than in other galaxies. The problem is simply that we are right among the stars of the Milky Way, and to find out where they lie we have to determine their distances. On the other hand, the shape and structure of distant galaxies is revealed by a glance at a photograph. It's essentially the difference between measuring the shape and extent of a forest by standing in it and determining the positions of individual trees, or flying over it in an aircraft. The concerted attack of modern astronomical techniques has, however, made considerable inroads into solving the problem.

It turns out that the Galaxy is flat and circular, with a bulge towards the centre – like two fried eggs stuck back to back. The Sun lies about two-thirds of the way out from the centre. All the stars making up the disc are revolving in circular orbits around the centre of the Galaxy, rather as the Earth revolves around the Sun, but on an enormously larger scale – and with the difference that there is no single, very massive body at the centre of the Galaxy controlling

the star orbits. Every star in the Galaxy feels the gravitational pull of every other star, and these attractions average out so that the overall pull on an outer star – the Sun, for example – has the same effect as would a large body at the Galaxy's centre. The Sun's orbit takes it around the Galaxy at a speed of 250 kilometres (155 miles) per second, but our Milky Way system is so vast that one round trip takes 225 million years.

Many of the brighter stars in the Galaxy's disc are concentrated into double spiral pattern. This discovery came as no surprise to astronomers, for most other flat galaxies have spiral patterns in the disc, marked out by bright stars. Indeed, galaxies with a flat disc are generally called spiral galaxies, in recognition of their beautiful shape. Photographs of these spectacular galaxies tend to overemphasize the importance of the spiral arms, for the majority of stars in the disc – dim, dwarf stars – are spread out much more evenly. Theoretical astronomy has recently shown that stars moving in a disc will tend to bunch together in spiral patterns, under the influence of their gravitational pull on each other. These calculations have been verified by computer simulations: a program specifies how each star in a disc of several thousand stars is moving initially, and the computer works out how the gravitational effect of each star on all the others affects their motions. Display the results as a film, and you can see the 'stars' almost miraculously bunch into a double spiral pattern, while still circling the 'galaxy'.

Although most of the stars we see lie in the Galaxy's disc, they are not the only members of the Milky Way Galaxy. All around the disc, 'above' and 'below', there are faint, old stars, arranged in a huge, spherical volume of space, the *halo*. Many of them are concentrated into dense globular clusters of stars, each containing up to a million members. These clusters, and the other halo stars, were the first stars to be born when our Galaxy formed, but now they are faint, senile relics of the Galaxy's past, slowly fading away while the disc remains young and vigorous.

Between the stars

The youthfulness of stars in the disc is due to one very important ingredient which has not been mentioned yet: the gas that lies between the stars. This interstellar gas, consisting mainly of hydrogen and helium, is exceedingly tenuous, far thinner than the best 'vacuum' ever attained on Earth. On average, there are half a dozen atoms in a matchbox-sized volume of space, but the Galaxy is so huge that the total amount of gas in it could make up 10,000 million Suns. And making stars is precisely what this gas is doing. Large, glowing clouds of gas – *nebulae* – which are denser than average, pull themselves together by gravitational attraction, and condense into a cluster of stars. Stars are forming all the time: astronomers have seen two faint stars 'switch on' within the last fifty years. The spiral arms are prime sites for star formation, because the pattern of gravitational fields here compresses gas into dense clouds.

In other galaxies, we can see clearly that the spiral arms are marked out by strings of nebulae, as well as by bright, newly formed stars. There's enough gas in the Milky Way Galaxy to create stars for thousands of millions of years to come, so the Galaxy's disc will long retain its youth.

Mixed in with the hydrogen and helium gas, other elements have formed into tiny solid grains. These 'dust' particles are about the size of the particles in cigarette smoke, and in a similar way they tend to block off light. The dust mixed with the general gas in the Galaxy's disc obscures distant regions; the light coming from the regions close to the centre of the Galaxy, for example, being dimmed some 100,000 million times by the dust along our line of sight. Fortunately, modern astronomers can use radiation emitted at infra-red and radio wavelengths to study such regions, for these rays can penetrate the dusty veil.

Planets

The dust is more obvious when it forms small dark clouds in front of the bright star-forming nebulae. These clouds appear in the shape of dark lanes, small round globules or 'elephant trunks', silhouetted against the glowing nebula background. It is from these dust grains, coming together as the gas collapses to form a star, that planets like Earth are made. The material making up all everyday objects – including our bodies – was once in the form of tiny interstellar grains, obscuring the view of any extra-terrestrial astronomer of several thousand million years ago.

In 1977, astronomers observing at infra-red wavelengths discovered a planetary system – another solar system – forming in a gas cloud in the constellation of Cygnus. The nature of the radiation suggests that the central star of this system is less than a thousand years old, and that planets are now condensing out of a surrounding disc of gas and dust. Our own solar system was born in this way some 4,600 million years ago, according to the age of some meteorites, the solar system's oldest rocks. Dust grains gradually built up into nine major planets and their moons, leaving a certain amount of rocky debris around the system. The planets furthest from the Sun, like massive Jupiter, pulled the surrounding hydrogen and helium gases to themselves by gravity, accumulating huge atmospheres until the planets became more atmosphere than rocky core. Thus the family of outer planets, the gas giants, was born. On the other hand, the inner planets, including Earth, were too close to the Sun, which then shone more fiercely than now, to hold on to any atmosphere. Any gases they may have collected were 'boiled off' by the Sun's heat, to leave the rocky worlds that we see today. Later on, gases escaping from volcanoes clothed Venus, Earth and Mars with an atmosphere, while the small bodies, Mercury and our Moon, had too little gravitational pull to retain this second atmosphere against the Sun's heat.

Our knowledge of the solar system has recently leapt forward with the use of space probes which can fly past a planet, orbit it, or even land on the surface. In the past two decades, our knowledge of the planets has virtually doubled. And by the end of this century, the inner planets should be as well studied as the Earth (in astronomical terms); while all the outer planets (apart from tiny Pluto) will have been

visited by probes, investigating them at close quarters. One very important consequence is that the study of 'comparative planetology' can tell us far more about the Earth. Its similarities with neighbouring planets will show what features are common in planetary formation and planetary geology, while the differences will highlight the reasons why the Earth has become what it is – in particular, why it seems to be the only planet which supports life.

Measuring distances

One very important prerequisite for planning a planetary probe mission is an accurate knowledge of the distance to the target planet. As a result, planetary distances must now be measured to a far greater accuracy than astronomy alone requires. Since the paths of the planets are controlled by the Sun's gravitational pull, the law of gravitation can actually reveal the relative distances of the planets very accu-

23

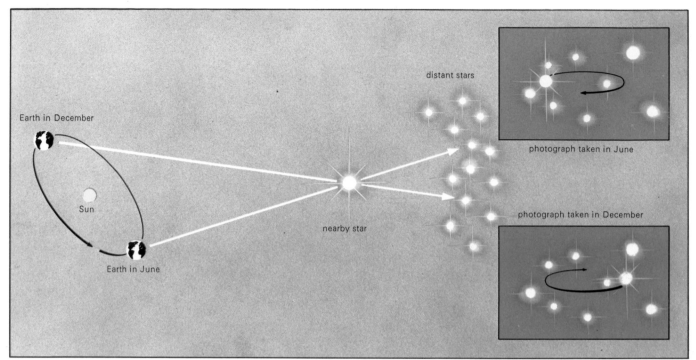

distant stars

photograph taken in June

photograph taken in December

nearby star

Earth in December

Sun

Earth in June

rately when their orbital periods ('years') are known. All distances in the solar system can therefore be calculated once just one distance has actually been measured. The method used today is radar – exactly the same in principle as ordinary aircraft radar. A radio wave transmitted from Earth is reflected off a planet, and the returning weak echo is picked up and amplified. It is possible to measure the delay time between transmission of the signal and return of the echo extremely accurately, and multiplying the delay time by the speed of light (the speed at which radio waves travel) gives the total distance covered by the radio wave – twice the distance to the planet. Venus is usually used as the target, because it approaches the Earth more closely than any other planet, and it has a comparatively large surface to reflect the radar signal. Despite the tremendous speed of the radio wave – 300,000 kilometres (186,282 miles) per second – it takes some five minutes over the return trip across interplanetary space.

From Venus' distance, and the 'scale model' of the solar system which follows from the laws of gravitation, the distance of the Earth from the Sun can be calculated. (The Sun itself cannot be used as a target for radar ranging because the intense radio emission from its atmosphere would overwhelm the weak reflected signal.) This fundamental distance, which astronomers call the *astronomical unit*, is probably the most accurately measured quantity in astronomy: it is 149,597,870 kilometres (92,750,679·40 miles). The scale of the solar system is now known with tremendous precision – equivalent to knowing the distance between the Eiffel Tower and Nelson's Column in London to within one centimetre (or less than half an inch).

But once we extend our scale to distances outside the solar system, we are lucky if distances are known to one part in a hundred. Sometimes astronomers are happy to know that the distance to a newly discovered galaxy, for example, is within 50% of the correct value. It is not obvious at first sight that the distances of the stars can be measured at all. Certainly radar is no help: the furthest contact yet made is with Saturn's rings, well within the solar system. But modern astronomers do have a dozen or so methods for gauging star distances. Some of them can only be used on specific types of object – clusters of stars, or pulsating stars, for example – but they all have their uses, and astronomers can cross-check the methods by giving the same distance to a pulsating star in a star cluster, for example.

The earliest method to be used, and still perhaps the most important, is *parallax*. We meet the principle of parallax all the time: hold a finger in front of your nose, shut each eye in turn, and the finger will seem to move back and forth against a distant background as your viewpoint changes. The further away the target is, the smaller is the apparent movement. Try 'winking' at a window frame across the room, seen against a distant house, and it will seem to move less than the finger did; if the target is a tree outside, silhouetted against the Moon for a background, the motion is hardly detectable. If you were to measure the angle which the target seems to move through, it is possible to calculate its distance by knowing the distance between your eyes.

At one time astronomers measured the distance to the Moon by parallax, by determining its precise position relative to the starry background simultaneously at two very widely separated observatories. Modern technology has now overtaken this approach, and the Moon's distance can be measured to an accuracy of 30 centimetres (12 inches) by timing laser pulses reflected off specially designed reflectors left on the lunar surface by manned and un-manned spacecraft. To measure the parallax of a star, how-ever, is a far more difficult procedure. The distance between two observatories on Earth is not nearly a long enough base line to produce an appreciable shift in a star's position, and astronomers have to make use of the Earth's orbit around the Sun.

An observatory takes a photograph of the suspected nearby star in, say, June, and another six months later in December. The Earth is now at the far side of its orbit, 300 million kilometres (186 million miles) from its position in June, and if the star is relatively nearby it will seem to have moved slightly when compared to the background stars. Another photograph taken the following June should show the star back in its original position, and confirm that the shift is due to parallax. By measuring the angle involved, and knowing the size of the Earth's orbit, the star's distance is easily calculated.

In practice, it's more complicated. Both the Sun and the star are moving in straight paths through space, and so the star won't return exactly to its original position. It's like watching someone walking through a fairground while you're riding on a merry-go-round. His position against the background is quite complicated, but you can separate his real motion from the back-and-forth movement which is due to your ever-changing viewpoint. Similarly, astrono-mers watch nearby stars year after year, and can eventually separate out the intrinsic movement ('proper motion') from the yearly 'wobble' due to parallax. And this wobble is incredibly small, even for the nearest star: if you were to blink your eyes at a tree silhouetted against the Moon, to get the same smallness of back-and-forth movement, it would have to be a tree ten kilometres (6·20 miles) away!

Patient observations – thousands upon thousands – at observatories all around the world have now produced fairly accurate distances for the thousand nearest stars. The nearest 'star' of all is actually a triple star system, Alpha Centauri. The two brighter stars are very close together, and make a bright southern-hemisphere star, known as Rigil Kent to air navigators. The third star is so faint that it is visible only with a telescope, and it lies slightly nearer to us. In view of its distinction of being the closest star known, this feeble little dwarf star has received its own name: Proxima Centauri.

When it comes to actually quoting star distances, the immensity of space makes the kilometre or the mile far too puny a unit. Even in the solar system, we are dealing with millions of kilometres, but Proxima Centauri is 40 million million kilometres from the Sun – while our Galaxy is a million million million kilometres across. In the face of such numbers, astronomers have chosen more convenient units, either based on the parallax method (using a unit called the *parsec*), or on the distance light travels in one year (where the unit is the light year). Both units of measure-ment are used in astronomy, often to the confusion of the casual reader; but we shall stick to light years here. Proxima Centauri is 4·2 light years away – a statement which not only tells us its distance, but also that the light we see from it now actually left it over four years ago. (For comparison, light from the Sun takes only $8\frac{1}{4}$ minutes to reach the Earth.)

The parallax method, even when used with great care, can only reach out to stars within 70 light years of the Sun. Beyond this, the tiny yearly shift becomes too small to measure. This volume of space, containing our thousand nearest stars, is, however, minuscule compared to the total 100,000-light-year extent of our Galaxy of 100,000 million stars. Fortunately, the other methods mentioned, par-ticularly that which applies to clusters of stars, can reach further out into the Galaxy. And once the distances to many stars have been ascertained, it becomes evident that despite the apparently enormous differences between the stars, there is an orderly system behind their properties.

Astrophysics: the natural history of stars

It is certainly a formidable task to try to interpret the shining, twinkling points of light in our night sky as Suns in their own right, and to calculate just what they would look like close up – and what is inside them. Early astro-nomers were content merely to measure the positions of stars; and interpretation did not really begin until a century ago, when astronomers first began to measure star distances. A new branch of astronomy sprang up, whose purpose was to deduce the properties of stars by applying the laws of physics to astronomical observations. It soon became known as *astrophysics*.

In the long run, astrophysicists want to know how stars are born, how they shine, how they change as they grow older, and how they eventually die. These questions can only be answered by beginning with the observations of stars as they are. Look carefully at the brighter stars on a clear night, and you'll notice that they are different colours. Some are red (Betelgeuse in Orion; Antares in Scorpius), others orange (Arcturus in Boötes), some yellowish (the Sun; Capella in Auriga), while there are also pure-white stars (Vega in Lyra) and bluish-white stars (Rigel in Orion). (On an average night you'll notice these stars

flashing other colours, too, but this is just an effect of the Earth's atmosphere, like the 'twinkling' of starlight.) The colour of a star is simply a measure of the temperature of its surface. Any object heated enough will radiate light, like a red-hot poker from a roaring fire; and the colour changes with temperature: red at low temperatures, changing to orange, yellow, white and finally blue at successively higher temperatures. Modern astronomers have instruments to measure colour with more accuracy than the human eye can achieve, but the principle of temperature measurement is just the same.

An even more obvious difference between the stars in the sky is in their apparent brightnesses. The Greeks originally divided the stars into six classes of brightness, so that the brightest stars are magnitude one, and the faintest visible to the unaided eye are magnitude six. Modern astronomical techniques can split these classes into much finer subdivisions than is possible with the eye, and large telescopes can reach far fainter stars. Because of lingering tradition, the fainter stars have higher magnitudes (23rd magnitude stars being about the faintest now observable with large telescopes), while very bright objects have negative magnitudes. So the brightest star in the sky, Sirius, has a magnitude of $-1 \cdot 4$. The Sun can be measured on the same system, incidentally, and turns out to have a magnitude of $-26 \cdot 8$; it's interesting to realize that the faintest stars astronomers can now detect are as much fainter than Sirius as Sirius appears fainter than the Sun! The wide range in

star magnitudes, from −1·4 to 23 (and fainter, presumably), is partly due to the differing distances of stars from us, and partly because stars do actually shine with different brilliances. Once a star's distance is known, it is easy to calculate its real output of light – its luminosity.

The brightest stars shine a million times more brightly than the Sun; but these celestial beacons are very rare: there are only half a dozen or so in an entire galaxy. Very faint stars are by far the most common variety, but it's obviously not as easy to find them in the crowded sky. The nearest star, Proxima Centauri, is just such a faint dwarf star, ten thousand times dimmer than the Sun, and it would have been overlooked had it not been so close.

With the first facts about a star – its surface temperature and its luminosity – well established, the astrophysicist can begin his investigation. There is a law of physics relating the brightness of any hot object to its temperature, and to its size. The radius of a star can then be calculated. Some stars turn out to be giants – Betelgeuse is four hundred times larger than the Sun – while others are dwarfs, no larger than a planet (but with the crucial difference that they are hot and shining in their own right, while a planet is cold and can only reflect light from a star).

When a physicist is faced with two measured quantities, like the temperatures and luminosities of stars, his first instinct is to plot a graph. A graph can often be confusing to anyone other than the person who plotted it, but a good graph can show up relationships which are hidden in a mass

The Hertzsprung-Russell diagram sorts stars according to their luminosity (intrinsic brightness) and temperature. The more luminous stars appear towards the top of the diagram; hot (blue) stars are on the left, and cool (red) stars to the right. Most stars fall near a diagonal line, the main sequence; a few large stars (giants) occur in the top right, and some very small stars (white dwarfs) to the lower left. This pattern in star properties allows astrophysicists to understand how stars shine, and how they should change over thousands of millions of years.

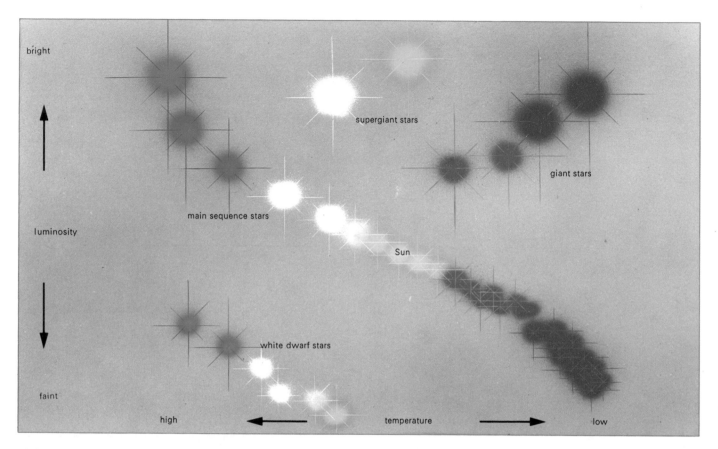

of figures in the original list of measured quantities. In a graph of star luminosities and temperatures, each star appears as a dot, whose position upwards denotes its luminosity, and whose distance sideways corresponds to its temperature. Significantly, the dots of most stars are found to lie along a narrow band on the graph. This main sequence of stars shows, in a graphical way, that stars progressively hotter than the Sun are progressively brighter, while cooler stars are progressively fainter. We shall see later that this uniformity is due to the way in which these stars shine. (Incidentally, the luminosity–temperature graph is often referred to as the 'Hertzsprung–Russell diagram' after the two astrophysicists who first used it in the early part of this century.

Down in the lower corner of this graph are points representing stars which are hot, yet very small and faint. These white dwarfs are the corpses of stars which have long ago burnt out, and just continue to shine dimly as they radiate their stored heat out into space. A white dwarf contains as much matter as the Sun, packed into a sphere a hundred times smaller. A matchbox of material from a white dwarf would weigh more than ten tons!

In the opposite corner of the luminosity–temperature graph lie the red giant and supergiant stars, hundreds of times bigger than the Sun. But a typical red giant contains

no more matter than the Sun – it is simply spread out over a huge volume. The edge of the star is so far out from the centre that gravity is very weak, and the outer layers of a red giant are continuously wafting off into space. Red giants are in fact ordinary stars passing through a temporary phase of swelling to giant proportions before dying as a tiny white dwarf.

Measuring the mass of stars – the quantity of material in them – is obviously very important in understanding their make-up. The only way of measuring a star's mass is by the effect of its gravitational pull; and in practice this means that we can only 'weigh up' stars which are in a double system, where two stars revolve around one another in orbits determined by the force of gravitation. Fortunately for astronomers, the majority of stars do occur as binaries – the Sun is in the minority here – and star masses are found to range from some twenty times lighter than the Sun to fifty times heavier. Going along the main sequence from bottom to top, we find that stars which are more luminous and hotter are also the more massive – further clues to the internal state of stars, which will become clear as we probe the secrets of main-sequence stars in Chapter 8. The astrophysicist displays this relationship on another graph, the mass-luminosity relation, in which the main-sequence stars again occupy a narrow band.

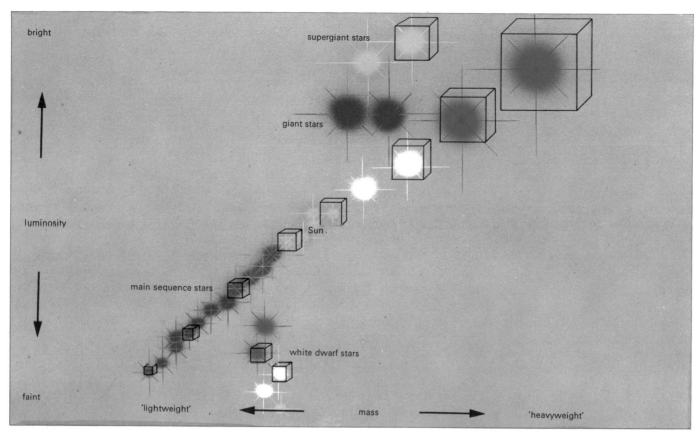

The nonconformist stars

These two graphs summarize very conveniently the observations of normal stars – it's a 'demographic breakdown' of the ordinary 'citizens' of the Galaxy. But the excitement and variety in life is often supplied by the bizarre, the nonconforming member of society, and our Galaxy has its fair share of such stars. We've already seen that red giants cannot control their outermost layers very effectively, and many of them pulsate slowly in and out over a period of months. Most famous of these is Mira (Latin for 'the Wonderful', so named by mediaeval astronomers) in the constellation Cetus. It is usually too faint to be seen with the unaided eye, but at maximum light it is one of the brightest stars in its region of the sky. The hotter yellow giants pulsate more rapidly and regularly; these are called *Cepheids*, and have achieved fame for their use in distance measurements outside our Galaxy.

Among the stars whose light seems to vary, some are shams. Their dimming is simply caused by one star in a binary system passing in front of the other and hiding its light. The most famous of these 'eclipsing binaries' is Algol, in the constellation Perseus. Its name means 'the Demon' in Arabic, and its position in the constellation represents the head of the monstrous Gorgon Medusa; so it's likely that the ancients knew of its regular dimming. However, the first good description of Algol's peculiar behaviour was by the deaf-mute British astronomer John Goodricke in the eighteenth century. Many binaries are, like Algol, too close for a telescope to 'split' them, and their double nature is determined by roundabout methods. When stars are extremely close, they are distorted by the tides they raise on each other, and each star is stretched out into an egg shape – if we could travel to the star Sheliak, in Lyra, we would see two stupendous shining eggs end to end, spinning around a point mid-way between!

When one star is smaller than the other, even more bizarre behaviour can occur. Gas pulled off the larger star forms into a rotating disc around the small star (a white dwarf, or an even smaller neutron star – see below) rather like an enormous version of Saturn's rings in gaseous form. The hot gas emits vast quantities of X-rays, which can be picked up by specially-designed 'X-ray astronomy' satellites. Occasionally, the gas falling on to the surface of a white dwarf will suddenly explode as a cosmic-scale H-bomb. The explosion shines thousands of times more brightly than the original star system, and we see a 'new star' – or *nova*, Latin for 'new' – appear in our skies.

Novae are not the most spectacular of nature's firework displays, however. Occasionally a star will blow itself apart entirely in a *supernova* explosion, becoming temporarily

29

thousands of millions of times brighter than the Sun. Such a cosmic cataclysm can, for a while, cause a star to outshine the whole galaxy of stars in which it has for long been an inconspicuous member. After a supernova explosion, the star's collapsed core may be left as a minuscule *neutron star*. Imagine the mass of the Sun compressed into a ball only a few kilometres across, and you'll gain some idea of the extraordinary density of a neutron star – a pinhead of material from it would weigh a million tons! Alternatively, the centre may fall together even more, resulting in that mysterious entity, a *black hole*. In this compact region of space, smaller than a neutron star, so much matter has been packed together that its gravitational pull has reached an irresistible strength. Not even light itself can escape from a black hole, and anything which falls in will be forever lost from the Universe. It is not yet certain whether black holes exist in the Universe – after all, it's impossible to actually *see* one, but there is more and more indirect evidence that this strange species does exist.

Around the explosion site of a supernova, the expanding gases blown out can make a beautiful, glowing nebula, lasting for thousands of years after the explosion itself. The Crab Nebula in Taurus is the most famous, a twisted filamentary structure left from a supernova which the Chinese saw explode in 1054 (although it's so far away that the actual explosion took place some six thousand years before that). In contrast, debris from the much weaker nova explosion soon disappears.

But another remarkable type of nebula surrounds some old stars. These are misleadingly called planetary nebulae (in fact, they have nothing to do with planets, but they do resemble planet-like discs seen through a small telescope): they are the outer layers of a red giant, gently puffed off into space, without all the fuss and fury of a nova or supernova explosion. Planetary nebulae are usually round, and have acquired names like the Ring, the Dumbbell, and the Owl Nebulae. Names like these date from the days of descriptive astronomy at the telescope: modern discoveries usually end up with a catalogue number, or at best, a name like 'the Red Rectangle'!

Probing other galaxies

The advantage of living right in the disc of our Galaxy is that, although we can't see at a glance how its contents are arranged, we can study them in great detail. The other galaxies, filling the Universe as far as the telescope can reach, seem to be made up of the same types of object as we see in our own Galaxy. Some are made of stars only, with no gas; while others have much more gas and dust than the Milky Way Galaxy. But our knowledge of the stars and the gas clouds in our own Galaxy gives us a good guide to what is happening in the distant galaxies in the Universe.

The distances to other galaxies can be measured with varying degrees of accuracy by several methods. Most rely on the fact that we know the actual brightness – the luminosity – of certain objects ('standard candles') in our own Galaxy, and when we identify such objects (such as Cepheid variables or nova explosions) in other galaxies, their relative dimness immediately tells of the galaxy's distance. It turns out that our Galaxy is not isolated in the Universe, but is part of a small group of galaxies. This *Local Group* includes the two nearest galaxies – the Magellanic Clouds – visible only in the southern hemisphere, and first described by Ferdinand Magellan on the first circumnavigation of the world (1519–22). The Andromeda galaxy, visible to the naked eye on a clear night, and the fainter Triangulum galaxy nearby, are also in the Local Group; and they are spirals, like the Milky Way. The Andromeda galaxy is actually the furthest object you can see with the unaided eye – it's $2\frac{1}{4}$ million light years away. The light we now see from this galaxy left it before there were humans on Earth.

Moving outwards into the depths of space, the realm of galaxies continues as far as the largest telescopes can see. Literally millions have been photographed. And these are just the brighter ones: there are many more galaxies too faint to be seen. There's a large cluster of galaxies in the constellation Virgo, some of whose members can be seen with binoculars. The light we see from them left at the time that the dinosaurs perished. On long-exposure photographs of the most distant galaxies, we are actually seeing them as they were before the Earth was born. Astronomers studying the furthest depths of the Universe are time travellers. Like it or not, they are looking at an earlier age of the Universe, and to compare these results with similar data from nearby galaxies may be misleading. Indeed, looking back in time like this, astronomers find more and more *quasars*, galaxies which are suffering an enormous central explosion. Quasars are probably very young galaxies, and we are looking out back into the 'maternity ward' of the Universe.

The 'Big Bang'

But the most surprising fact about galaxies is simply that they are all moving apart. The usual interpretation is that the entire Universe is expanding and carrying the galaxies with it, like raisins in a fruit-cake expanding in the oven. The observations certainly support this interpretation, and with it the implication that the Universe must have been 'smaller' in the past. Indeed, it seems that some 15,000 million years ago, the whole Universe was packed together into a small, unbelievably dense ball, which exploded. The continuing expansion from this initial 'Big Bang' is still evident in the galaxies' racing apart. Fantastic as this idea is, astronomers have found that it fits in with many apparently unrelated astronomical facts: the ages of stars; the abundance of some elements in space; and even with a faint background of radio waves probably lingering on from the inferno of the Big Bang.

Cosmologists studying the early history of the Universe are trying to push back in time, to find out just what went on in the first fraction of a second after the Big Bang. The problem here is that the matter was so dense – more closely packed than in a white dwarf star, denser even than the matter of a neutron star. In those conditions, matter as we know it could not exist. Atoms certainly could not: there would be 'free' electrons, protons and neutrons. But in the conditions which cosmologists are investigating, even these particles are not stable: matter would be made of – what? The only hope of answering this question lies with the experiments of particle physicists in Earth-based laboratories, as they peer ever deeper into the successively smaller particles making up matter. One day they should be able to say what matter was like in the conditions of the Big Bang.

And so the wheel comes full circle. On our tiny speck, Earth, a busy little species called *Homo sapiens* is building complex pieces of equipment to look right into the heart of matter. His object is simply to find out what everything is ultimately made of. But as astronomers peer outwards from our planetary system, through the stars of our Galaxy, and out into the reaches of intergalactic space, they find that the Universe itself began in conditions much closer to that of a target in a particle physics experiment than the emptiness of the vast, lonely reaches which astronomers usually associate with the Universe. By investigating fundamental particles in a laboratory, the physicist is helping the astrophysicist to explain how, from a colossal Big Bang some 15,000 million years ago, our Universe became the astonishing expanse that it is today.

A beam of electrons is confined by magnetic fields to follow a narrow track through a vacuum chamber. Such tests should lead to the control of the very hot 'gas' called plasma, and possibly to generation of vast quantities of energy by the fusion of hydrogen atom nuclei.

From the physicists' point of view, the Universe is, as already mentioned, built up from just a few types of fundamental particles: the members of the quark and lepton families. All the diversity of our surroundings, and the stupendous wonders in the depths of space, are merely the result of arranging Nature's basic ingredients in different ways, under the influence of the forces which they exert on each other.

The diverse family of leptons

We have already had a brief introduction to the lepton family. For practical purposes the most important by far is the busy electron. Whether it is whirling around the nucleus of an atom, moving through a wire as an electric current, joining atoms together as molecules, or speeding through the vacuum of a television tube to create a picture, the electron is indispensable to us. It is a fairly insubstantial little particle, 1,836 times lighter than the proton, for example (*leptos* is the Greek for 'small').

One electron carries one unit of negative electric charge. In everyday life, we think of electric charge (on a rubbed comb, for instance) as something we can change to any value we like. Nature has decreed, however, that charge comes only in packets of a certain size – and each electron carries just one of these basic units of charge.

In modern physics, the idea of a basic unit of something – a smallest possible 'packet', or *quantum* – is quite natural. Since the quantum theory revolutionized physics at the beginning of this century, scientists have come to accept that most quantities, including energy, come in discrete 'packets' – that is, they consist of single units and can only be increased or decreased stepwise, in steps of one unit. Another property of particles, for example, *spin* – the degree of angular momentum possessed by a particle – is also quantized in this way. But everything in our large-scale world is so vast in comparison with these basic units that changes seem to be continuous. Switch off a record-player turntable, and it appears to slow down gradually as friction brings it to rest. In fact, however, it is losing whole units of spin, one at a time, and slowing down step by step. But it begins with such a huge number of spin units that the spinning turntable is losing some ten million million million million million units every second – so the appearance of a continuous slowing-down is hardly surprising.

Only very sensitive experiments can detect the changes of a single step in Nature's units. The celebrated American physicist Robert Millikan first measured the tiny electric charge of single electrons in 1912, by balancing small electrically charged oil drops between two forces – the downward pull of gravity, and an applied electric field pulling the negatively charged drops upwards. He watched the drop through a microscope, and when the critical balance was reached the oil drop hung motionless in mid-air. Millikan could then calculate the charge on the drop: it always turned out to be an exact multiple of one particular figure, because each drop was carrying a certain number of extra electrons.

Actually performing the experiment wasn't quite this easy. While he was watching a drop, its charge could change by one unit if it lost an electron, or gained one from its surroundings. Its critical balance now upset, the drop would begin to drift up or down, and Millikan quickly had to readjust the electric field. But then the differently charged drop gave him another result; and so following the fortunes of a single drop was the most efficient way of carrying on the experiment. On one occasion, Millikan peered through his microscope at one tiny oil drop for a continuous stretch of eighteen hours!

The electron is the most studied of the lepton family, on account of its importance in making up matter. Its big brother, the muon, has always been more of an enigma. First discovered in 1937, while physicists were searching for another particle, its nature was misconstrued for many years ('muon' is short for 'mu-meson', since at first it was classified as a *meson*, a 'medium-sized' particle, from the Greek *mesos*, 'middle'). With the recognition that it was a lepton, in spite of being 207 times heavier than the electron, physicists found its existence rather embarrassing. Why should Nature 'need' another lepton, particularly an unstable one like the muon, which lives for only a couple of millionths of a second before breaking up into an electron and two neutrinos? Modern theories suggest that the number of lepton types is related to the number of different quarks, and it is certainly interesting that at the same time as very recent experiments have begun to show evidence for a fifth quark, other researchers have found suggestions of another lepton, even heavier than the muon. This extremely short-lived lepton appears to be some 3,500 times heavier than the electron.

Completing the lepton family (at present) are the two known types of neutrino. There is one associated with the electron, and one with the muon (but presumably there is another associated with the new, heavier lepton). Neutrinos are the oddest particles discovered by particle physicists. They have no mass at all, and no electric charge, and they always move at the speed of light. The first and last points are actually related, for Einstein's famous theory of relativity prohibits any particle with mass from travelling at the speed of light; and conversely this involves the assumption that a massless particle (like a neutrino) can *only* move at the speed of light. In this case, the theory shows that the particle will carry energy with it, despite its lack of mass.

So neutrinos have energy, but even so they are remarkably elusive. They don't respond to either of the two strongest

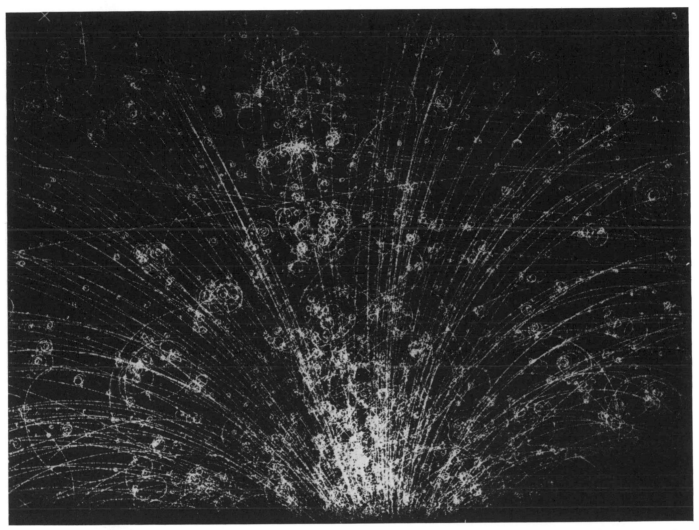

forces (strong nuclear and electromagnetic), and they only show up when they – very occasionally – react with another lepton in respect of the weak force. Neutrinos are very specific in their affinities, too. An electron neutrino will only involve itself with electrons, and a muon neutrino with a muón. For example, when a muon decays into an electron, there must be two neutrinos produced: a muon neutrino, associated with the original muon, and an electron neutrino as a companion to the resulting electron. (Strictly speaking, the latter is an electron *anti*neutrino, as we shall see later.)

The lepton family is quite a mixed bag. It's possible to get some 'feel' for what an electron or a muon is like, but the neutrinos seem to defy common sense. Let us accept these wayward members as Nature has given them to us, however, and meet the other family of particles.

Quarks: the nearly identical quads
The quarks seem a much neater group on first acquaintance. The four quarks don't seem to differ very much from one another – just enough to distinguish between them. This fact may actually result from our inability to take a look at isolated quarks. Unlike leptons, quarks are a tremendously gregarious family, bound together by the strong force, and physicists have not yet succeeded in breaking up a composite particle (such as a proton) into its constituent quarks. What we know about quarks must be deduced from how they behave in their tight little groups.

The names given to the quarks are less 'classical' than those of the leptons, reflecting the less traditional outlook of physicists in the 1960s and '70s. The four established quarks are called *up*, *down*, *strange* and *charmed*. The first two are 'everyday' quarks: they make up the protons and neutrons which form the nuclei of ordinary atoms, and they differ from each other basically in electric charge. The strange quark has an additional property, so baffling when physicists first came across it that they simply called it 'strangeness'. The strange quark can change spontaneously into an up or a down quark under the influence of the weak

35

force, but not through that of the strong force. The lifetimes of particles containing a strange quark were therefore much longer than expected – a 'long' time in this context meaning a thousand millionth of a second!

Similarly, the fourth quark has another property affecting its decay into other quarks. But by the time this quark was discovered in 1974, theory had already predicted its existence, and the rather less sensational name of 'charm' was attached to the new property.

(The naming game continues, with the likelihood of there being two more quarks. The fifth and sixth possible quarks have competing choices: either *beauty* and *truth*, or *bottom* and *top*. It remains to be seen whether physicists of the present generation will prefer the whimsical approach or the down-to-earth.)

The oddest fact about quarks, apart from the apparent impossibility of isolating individuals, is that their electric charges are not whole units of charge: the quantity of charge is either $\frac{1}{3}$ or $\frac{2}{3}$ that of the electron. But when other particles are made up from quarks, the total charge always works out as 0, 1 or 2 basic (electron) units. Present-day experiments can only study particles made up from quarks, and not the individual quarks themselves, and so all the particles which are actually detected in experiments do have charges based on the same unit as the electron. For example, the proton is made of two up quarks (each with charge

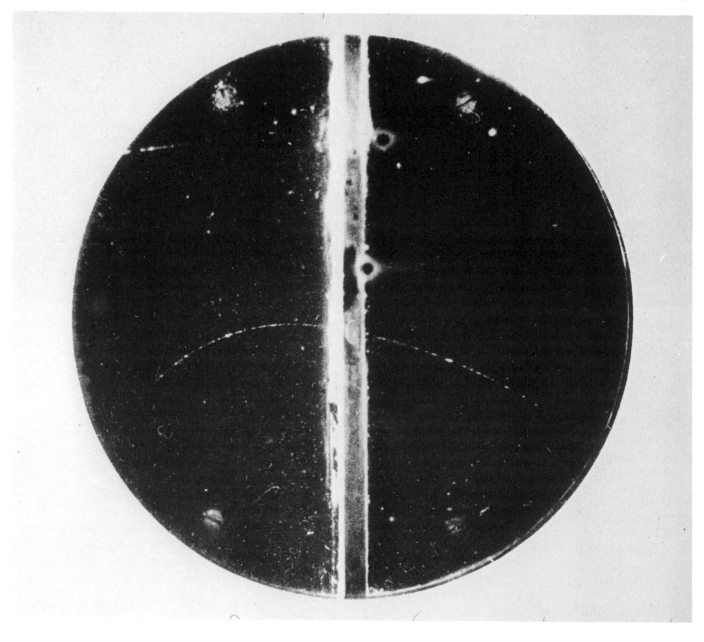

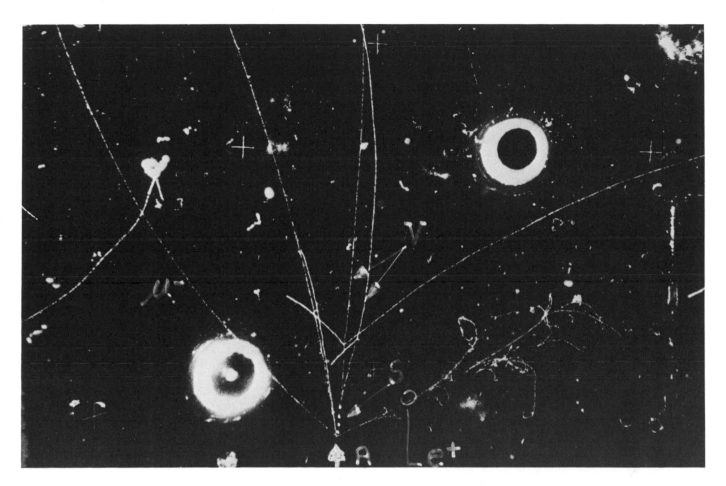

$+\frac{2}{3}$) and one down quark (charge $-\frac{1}{3}$), so the total charge works out as $\frac{2}{3} + \frac{2}{3} - \frac{1}{3} = +1$ unit of charge. In the neutron, the charges on two down quarks (each $-\frac{1}{3}$) cancel out the charge of one up quark ($+\frac{2}{3}$) to leave it with zero charge overall.

Other 'heavy' particles (*baryons*, from the Greek *barys*, 'heavy') can be made by taking different combinations of three quarks, and sticking them together with the strong nuclear force. Combinations containing one (or more) strange quarks are the 'strange particles', which puzzled particle physicists in the 1950s by their unusual way of disintegrating. One of the earliest successes of the quark theory was the prediction of a baryon consisting of three strange quarks: it would have a strangeness of minus three, and an electric charge of minus one. Just such a particle, called the *omega minus*, with the predicted mass, was found in 1964, and crowned the efforts of Murray Gell-Mann and his collaborators in sorting out the long list of the old 'fundamental' particles in terms of a simple internal structure of quarks.

Baryons containing the fourth type, charmed quarks, must also exist, although in the few years which have elapsed since the 'discovery' of charm, only one or two of these elusive charmed particles have yet turned up.

Despite Nature's generosity in providing eight ultimate particles, four quarks and four leptons (possibly to be increased to a total of twelve, if recent discoveries are confirmed), ordinary matter only utilizes three or four: the up and down quarks, and the electron – with the electron neutrino making a very occasional contribution. Up and down quarks combine to make the protons and neutrons in atomic nuclei, while the electrons may orbit around to build the atoms and molecules of everyday matter, or they may pursue their own paths, in the extreme conditions found in the far reaches of the Universe.

What need is there then for the strange and the charmed quarks, and for the muon – not to mention the new quarks and leptons? These particles are all unstable, breaking down within a fraction of a second. We can only study them by producing them in a modern particle accelerator, and recording their brief moment of existence. But these ephemeral particles hold important clues to the workings of the strong and weak nuclear forces.

The unification of the electromagnetic and weak forces by Steven Weinberg and Abdus Salam, for example, could not work if only three quarks existed, as was thought at the time. The bold prescription of a fourth quark, with a new property (charm), could solve the theoretical problem – and, indeed, particles containing a charmed quark were later found. Theorists now believe that the different types of quarks and leptons will be revealed as a by-product of the way in which the forces of Nature are related. Such a synthesis, relating the fundamental particles to the different ways in which a single unified force produces the effects of the four 'standard' forces, is one of the most important goals of modern physics.

The mirror world of antimatter

Right down at the level of the fundamental particles, physicists have discovered a whole new realm of existence: the mysterious world of *antimatter*. Each of the fundamental particle types has an antiparticle, exactly opposite to it in all its properties.

Take the electron, with its electric charge of -1 unit. The antielectron has a charge of $+1$. Or consider the recently discovered charmed quark, which has an electric charge of $+\frac{2}{3}$ and a charm of $+1$ unit; the charmed antiquark has a charge of $-\frac{2}{3}$ and a charm of -1. There is an antiquark for each type of quark; an antilepton for each type of lepton; and each antiparticle mirrors the properties of the corresponding type of particle. Shouldn't we then extend the list of fundamental particles, then, to include the antiparticles? We might think that our balance sheet of four quarks and four leptons should be doubled in size to accommodate four antiquarks and four antileptons.

Fortunately for the concept of the basic simplicity of the Universe, we don't have to do this. Each antiparticle is so exactly an opposite of the corresponding particle that it is evidently just the 'other side' of our familiar particles.

On this basis, the great British physicist Paul Dirac actually predicted that antimatter should exist some fifty years ago, four years before the first antiparticle turned up in experiments. He had combined Einstein's theory of relativity with the quantum theory, which described how particles behave on a very small scale; and to his surprise he found that it was quite possible for particles exactly opposite to the ordinary ones to exist. (Mathematically it was like solving a quadratic equation, which has one positive and one negative solution.) His ideas were at first treated as an interesting mathematical result, without any real meaning;

Albert Einstein first realized that matter can be regarded as a form of energy, and so opened the way to the discovery of

antimatter. Here Einstein is holding an informal seminar.

but the discovery in 1932 of the antielectron (usually called the positron, because of its positive electric charge), and later of other antiparticles, made it clear that antiparticles are definitely real. Indeed, present-day theories of the fundamental particles just will not work unless antiparticles are included.

The problem in studying antiparticles is that they are so exactly 'anti'. All their properties precisely mirror the appropriate particle: in electric charge, strangeness and charm, as we have already seen, and also in various other properties which physicists can ascribe to them. So, when an antiparticle produced in a laboratory meets a corresponding particle of 'ordinary' matter, every single property it possesses is cancelled out. A charmed antiquark collides with a charmed quark, for example: the first particle's charm of -1 is destroyed by the charm of $+1$ carried by the other; while its electric charge of $-\frac{2}{3}$ is annihilated by the $+\frac{2}{3}$ charge of the ordinary quark. Every property which distinguishes one particle from another is destroyed – with one exception: its mass.

Mass, in this sense, isn't a word we use a lot in everyday life. Roughly speaking, it's the 'quantity of matter a body contains' and it has two effects. One is gravitation. The gravitational force pulling two bodies together depends on their masses. An individual adult human is much more massive than a cat, so the Earth pulls the human downwards with a greater force – as we can demonstrate with bathroom scales. But suppose we took the scales to the Moon. It's so much less massive than the Earth that the pull of gravity between our bodies and the Moon is only one-sixth of that of the Earth in your bathroom at home, and the scales would show our weight to be proportionately smaller.

This doesn't mean that we are any slimmer simply because we are on the Moon: the amount of matter in our bodies – the mass – is the same. Using the force of gravity is evidently not the best way to measure mass, and scientists prefer to define mass in a different way: in terms of *inertia* – resistance to acceleration. Kick a football and a table-tennis ball with equal force. The latter shoots off much more readily, because it has the smaller mass. Even in the weightlessness of space, you would find exactly the same effect, because mass is one of the fundamental properties of an object, wherever it may be.

One apparently odd point about antimatter is that (despite its generally 'anti' properties) it has the same mass as its corresponding particle – it does not have a negative

39

mass. If we could make a football of antimatter, it would fall downwards under Earth's gravitational pull at exactly the same rate as an ordinary football. But there's one major problem with kicking an 'antifootball': the explosive annihilation of particles (in your shoe) with antiparticles (in the antifootball) would wipe out shoe, playing field and the surrounding town in a cataclysm of H-bomb proportions. (In fact, this would have occurred the instant that the antifootball was exposed to the real world.)

Albert Einstein first showed, back in 1907, that it is theoretically possible to convert mass into energy – and a stupendous amount of energy, too. To calculate it, multiply the mass involved by the speed of light, and then by the speed of light again: Energy (E) = mass (m) × speed of light (c) × speed of light (c); or in symbols $E = m \times c \times c$, which can be shortened to the most famous equation in science: $E = mc^2$. Einstein's equation shows that theoretically we could destroy just one kilogramme of matter of any kind to produce enough energy to run a giant power station for a year, replacing literally millions of tons of coal or oil.

Fortunately, a one-kilogramme pack of sugar will not spontaneously convert itself into energy in this way. The ultimate particles making up the sugar – up and down quarks, and electrons – all have electric charge and other properties which cannot just disappear, to leave 'pure mass' which could become 'pure energy'. Ordinary matter by itself is thus immune to total destruction. But we have seen that when a particle and its antiparticle are brought together, all the 'conserved' properties (like electric charge) cancel out completely – but both particle and antiparticle have positive mass, and all of this changes in a blinding instant into energy.

Physicists have difficulty in studying antiparticles for this very reason. It is not possible to collect them in any way, for as soon as they meet real matter they are destroyed. But by reversing the process, physicists can create antiparticles. When enough energy is concentrated at one point, it is possible that it will spontaneously turn into mass. It would not be possible to create just a single particle – say, an electron – in this way, for that would involve creating electric charge from nothing. But the simultaneous creation of an electron and its antiparticle, a positron, can occur. Since one has a negative, and the other a positive charge, we haven't added any overall charge to the Universe. 'Pair production' of particles and antiparticles is a common procedure in the present high-energy particle accelerators, and it is indeed the only way of 'making' antiparticles.

Just like ordinary quarks, antiquarks have not been found individually, but only bound up in heavier particles. Groups of three antiquarks make up the antibaryons – like the antiproton (2 antiup, 1 antidown) with a negative electric charge, and the antineutron (1 antiup, 2 antidown)

They also allow the experimenter to choose the type of which has no electric charge. Superficially, the neutron and antineutron seem very similar, with the same mass and no charge. But the quark constitution of the ordinary neutron (1 up, 2 down) is clearly different from that of the antineutron, and it is evident that a neutron meeting a particle of identical mass and no charge could 'recognize' from its quarks whether it is neutron or antineutron: friend or foe.

The banding together of quark threesomes is quite exclusive, in that we do not find mixtures of quarks and antiquarks in the same baryon. Obviously such an arrangement could not be permanent; but then the proton is the only long-lived baryon, and the others are only short-lived particles anyway, ultimately turning into protons. On the other hand, it is possible for one quark and one antiquark (either of the same or of different types) to form a short-lived pair. These combinations of quark and antiquark are called *mesons*, and they formed a separate family under the old classification scheme (which had to grow more and more complicated until it was understood in terms of just a few quarks). The name meson was applied to medium-mass particles (from the Greek *mesos*, 'middle'), as opposed to the more massive baryons (*barys*, 'heavy') and the lighter leptons (*leptos*, 'small'). These terms have now changed their meanings, and a baryon is now defined as a particle composed of three quarks, and a meson as a quark–antiquark pair; while the leptons are a separate family, distinct from the quarks. As a result, the masses of the different classes now overlap, totally confusing their original meanings: the newly discovered 'fifth lepton', for example, is twice as massive as the proton, the lightest of the baryons.

The commonest meson is the *pion*, a particle we shall meet again in connection with the strong nuclear force. It was predicted long before its discovery, and although the search was confused for a while by the discovery of the muon (which had a mass close to the prediction), the pion was successfully detected in 1947. In these early experiments, the particle accelerators available were not powerful enough to produce particles of this kind, and experimenters were forced to rely on Nature's own titanic particle accelerators – exploding stars. In these rare and stupendous supernova outbursts, protons are accelerated to enormous speeds, and after a lengthy coast through space, some of them crash into the top of the Earth's atmosphere. The nuclei of atoms in our air bear the brunt of the onslaught, and in the high-energy collisions many new particles and antiparticles are born. Positrons, muons and pions were all first found in the debris of these 'cosmic-ray showers'.

Even in modern particle accelerators we cannot reach the energies of the fastest cosmic-ray protons; but these machines do have the great advantage of producing a much larger concentration of particles in the fast-moving beam.

Cosmic ray particles are constantly impinging on the top of the atmosphere; balloon-borne experiments can detect a few particles far more energetic than any in the physicist's

particle accelerator. Because the electrically charged particles are deflected by the Earth's magnetic field, cosmic rays are concentrated towards the north and south polar regions.

They also allow the experimenter to choose the type of particle he sends through the accelerator, and to change the target. It's like using a torch to find your way around on a dark night, rather than having to wait for much brighter, but very infrequent, flashes of lightning.

Before leaving the realm of antiparticles, let's look at the possibility of finding antimatter on a larger scale. Theoretically, there is no reason why antiprotons and antineutrons should not stick together by the strong force to make negatively charged antinuclei. Positrons could then circle these antinuclei, and constitute antiatoms of the various elements. Chemically, antiatoms would react with each other exactly as the corresponding atoms do, and could make up antimatter on the macroscopic (everyday-sized) scale. Antiwater, antisand, even antihumans are possible: and they would live on an antiplanet, circling an antistar.

Are there actually such antistars, and antiplanetary systems, in the Universe? Certainly they cannot exist in our Galaxy, because they would be gradually annihilated by the tenuous gas between the stars. The energy from annihilation would be carried away by gamma-rays; and these rays (not to be confused with 'cosmic rays', which are actually particles, mainly protons) would have been picked up by the purpose-built detecting satellites.

But let's take the argument one step further. Although our Galaxy – planets, stars and interstellar gas – must be entirely ordinary matter, could other galaxies be made completely of antimatter? Light emitted by antimatter would be exactly the same as light from matter, so it is not possible to tell by looking at a galaxy, or even by analyzing its light in detail. There is, however, extremely tenuous gas filling the space between the galaxies, and again we should expect to pick up gamma-rays from galaxies which have strayed into regions of antigas, or antigalaxies surrounded by ordinary gas. Because these gamma-rays are not detected, astronomers generally agree that our Universe must consist entirely of ordinary matter: antimatter is a laboratory curiosity, created only to quickly disappear again in an annihilation with a particle of ordinary matter.

Nuclei: the key to diversity

Looking at the Universe as a whole, the vast majority of matter consists of protons and electrons. Since they occur in equal numbers (cancelling out the overall electric charge)

the heavier protons contain most of the mass of the Universe. This situation is not surprising, for protons and electrons are the only stable fundamental particles (apart from the neutrinos), and even if the Universe had been filled with a mixture of particles at an earlier age, as modern theories suggest, all the others would have decayed by now. The muons and other possible leptons would quickly have degenerated to electrons; while the quarks making up the baryons would have progressively changed to the lowest 'energy state', two up quarks and one down quark, which characterizes the proton. (We'll turn to the question of energy states later in this chapter.)

Even the neutron is not intrinsically a stable particle.

After an average lifetime of a quarter of an hour – an enormously long time in terms of other particles' life expectancies – one of the down quarks will spontaneously change into an up quark, altering the particle's configuration from one up quark and two down to two up and one down, the quark state of a proton. In the process, it shoots out an electron, which carries away a unit of negative charge to balance the positive charge on the newly created proton, and an electron antineutrino.

So why is the Universe not filled with only protons and electrons? (In chemical terms, this would constitute hydrogen gas – see below.) In that case the Universe would be a pretty dull place, with no planets and no life. The redeeming feature is that neutrons can be stabilized – prevented from disintegrating – by combining them with protons in tight little clusters bound by the strong nuclear force. A cluster like this forms the central, positively charged nucleus of all atoms other than hydrogen (whose central positive charge is carried by a single proton).

In the wider Universe outside the Earth, and in the controlled action of a particle physicist's accelerator, the nuclei can occur without the surrounding electrons which make up the atoms and determine their chemical properties. A nucleus is usually named after the atom it would form, however, even when it's in a situation where all its electrons have been stripped away. In the incandescent heart of a star, for example, three tiny clusters each containing two protons and two neutrons can fuse together to make a cluster of six protons and six neutrons. The original clusters are identical to the nuclei found in atoms of helium gas, while the product is the same as a carbon atom's nucleus. Hence the reaction can be written 3 He→C; and the symbols refer to the proton-neutron clusters, not necessarily in the centres of atoms at the time.

To find out why neutrons are stable in clusters (nuclei) but not outside, we must first ask why 'free' neutrons are unstable. It's basically a question of energy. Put very simply, whenever energy is cooped up in any way, it tries to spread itself out. The Sun is producing a vast amount of energy every second, because of the nuclear reactions going on at its core; and the heat and light radiation from its surface are spreading this energy out over the Universe. Or to take another example, the air under pressure inside a blown-up balloon has excess energy, and as soon as you release the neck the escaping air molecules spread out this excess energy throughout the room. Similarly, the quark configuration of the neutron makes it a higher 'energy state' than the proton, and it can lose energy by converting a down into an up quark. The extra energy is carried away and spread out as the energy of motion of the electron and the antineutrino, and also as the mass of the electron (according to Einstein's mass–energy equation).

This evening-out of energy, incidentally, is known as the Second Law of Thermodynamics, and it was first investigated in connection with different types of engines. Why shouldn't an ocean liner be able to pump on board sea water, extract heat energy from it to drive its engines, and put the cooler water back into the ocean? This kind of process is never possible, however, because it would mean a spontaneous concentration of energy from a spread-out source into one place. According to the Second Law, only processes producing a spreading-out of energy can occur spontaneously – you would never expect the air in a room to suddenly rush into a balloon and blow it up of its own accord. Taking a very long-term view of the Universe, the Second Law predicts that eventually all energy will be equally spread out: nowhere will be hotter or cooler than anywhere else, so there will be no way of generating power, and the Universe will end in a 'heat death'.

The concept of *entropy* is another means of expressing the Second Law. The entropy of a system is a measure of its disorder; in other words, energy always changes from a useable state to a non-useable state. When we burn coal in an electricity generating station, we are turning the latent energy in the coal into energy in the form of electricity, but the result is a net loss of useable energy: the coal is used up, and its latent energy cannot be recovered. The most interesting thing about entropy is that it determines the direction of time: time always travels in the direction of increasing entropy. If we made a film of a pendulum swinging, it would make no difference whether we ran the film backwards or forwards; but if we made a film of a pendulum slowing down as a result of friction, it would be immediately obvious which way the film should be run.

Returning to the neutron, we can now see how to prevent it from decaying: put it in a situation where its decay would need to *take in* energy. Two protons and two neutrons bound together by the strong force make up a helium nucleus. If one of the neutrons were now to decay, the cluster would contain three protons and one neutron, and this turns out to be a higher-energy grouping than the original helium nucleus. Although the nuclear binding force is essentially unchanged, there are now three, instead of two, positively charged protons repelling each other by the simple repulsion of 'like' electric charges. So while the decay of a neutron would give the neutron itself a lower energy, it would leave the nucleus in a higher energy state. Totalling up the energy credit and debit columns, we find that such a decay would require an energy input to the nucleus; the Second Law of Thermodynamics thus forbids the neutrons from changing once they are locked up in a helium nucleus.

Similarly, any other small cluster of protons and neutrons is stable, as long as the numbers of protons and neutrons are roughly equal. When these nuclei are clothed in electrons, they make up the atoms of the chemical elements, such as carbon, oxygen or iron. The chemical properties of the elements depend only on the number of orbiting electrons; and to ensure that the atom is electrically neutral as a whole, this must be equal to the number of positively charged protons in the nucleus. (This is how we can characterize a nucleus by the name of an element, even though the term 'element', strictly speaking, refers to chemical behaviour.)

Notice that the number of neutrons doesn't come into the argument at all, so it is quite possible to have two atoms with identical chemical properties, but with different numbers of neutrons in the nucleus. All atoms of carbon, for example, have six electrons in orbit, and six protons in the nucleus. Almost all carbon atoms also have six neutrons in the nucleus, but one in a hundred has seven. These heavier atoms behave chemically like ordinary carbon atoms and only sensitive physical measurements of atomic masses can tell them apart. Nuclei with the same number of protons (and hence the same chemical name) but different numbers of neutrons are called *isotopes* (from the Greek for 'equal place' – that is, their place in the chemical table of elements); and we can distinguish them by putting the total number of nuclear particles after the chemical name. Hence ordinary carbon is carbon-12, while the heavier isotope is carbon-13.

Isotopes are often most useful to us when their nuclei are not stable. These radioactive isotopes ('radioisotopes') behave chemically like normal elements, but they can be detected by the radiations they give off as their nuclei spontaneously change to more stable varieties. Medicine has many uses for radioisotopes: for example, the activity of the thyroid gland, which accumulates iodine, can be monitored by injecting a small quantity of a radioisotope of iodine into the bloodstream, and monitoring the build-up of radioactivity in the gland.

Practical physics and chemistry have also benefitted in countless ways from the use of radioisotopes; and their application has even reached the field of archaeology. The archaeologists' isotope is carbon-14, an unstable nucleus created by the impact of cosmic ray particles from space on the nuclei of carbon atoms in the carbon dioxide gas in Earth's upper atmosphere. As a result, the air contains a fairly constant, though very small, percentage of carbon-14 in the molecules of its carbon dioxide gas, as newly formed carbon-14 nuclei filter down to replenish the stocks, which constantly decay. The carbon-14 nucleus has too many neutrons (eight) for the six protons, and one neutron eventually decays to leave the nucleus with seven protons and seven neutrons: the stable configuration of nitrogen-14.

The American chemist Willard Libby pointed out that the naturally-occurring carbon-14 could allow archaeologists to date remains containing carbon – principally wood, bones and ashes. All the time a plant or animal is alive, it is exchanging carbon with its surroundings, and the proportion of the unstable carbon isotope stays the same as it is in the air. But as soon as the organism dies, and stops taking in fresh carbon-14, there is no replacement for the disintegrating carbon-14 in the remains, and the proportion gradually falls. Since the rate at which carbon-14 decays into nitrogen-14 is known, a scientist can measure the proportion of carbon-14 in a small organic sample (20 grammes is enough for an analysis) to calculate how long ago death occurred. Dates measured this way extend back 60,000 years into Man's past.

It took archaeologists ten years, however, to accept fully that the young science of nuclear physics could help their field of research. Indeed, their final acceptance of carbon-14 dates involved a complete rethink on European prehistory, for many of the megalithic monuments (such as Stonehenge) turned out to be associated with wood and animal remains a thousand years older than the dates which archaeologists had ascribed from the slender evidence they had before. Archaeologists now accept that builders of these monuments were an earlier, highly advanced people of north-west Europe, rather than, as originally thought, a later culture which crudely copied the magnificent temples of the eastern Mediterranean peoples.

The unstable giant nuclei

Moving on to larger nuclei, we find that with increasing size, the number of neutrons becomes steadily larger than the number of protons. The most common isotope of oxygen, for example, has 8 protons and 8 neutrons; but iron has 26 protons and 30 neutrons; and the most common lead isotope contains 82 protons and 126 neutrons. The imbalance is caused by the battle between the strong and the electromagnetic forces, which we have met already in the helium nucleus.

The electromagnetic force can be felt over a long range, so every proton feels an electrical repulsion from every other proton in the same nucleus. But the strong nuclear force, binding the nucleus together, acts over only an extremely short range, and a proton feels an attractive force only from its nearest neighbouring protons and neutrons. This doesn't matter for the smaller nuclei, since all the particles are close together, but larger nuclei must contain a higher proportion of electrically neutral neutrons to provide more binding force without adding any more positive charge.

But for nuclei larger than lead, even the extra neutrons cannot help permanently, and sooner or later these overweight nuclei break up in a process of *radioactive decay*. This radioactivity is far more spectacular than that of a small nucleus like carbon-14, for the powerful repulsion between the protons tears off whole chunks from the nucleus, each chunk consisting of a helium nucleus: two protons and two neutrons.

A grotesquely overweight nucleus like uranium-238 will lose eight helium nuclei in succession before it reaches a stable state, turning into lead. One early use of this continuous stream of high-speed particles from heavy nuclei was to provide the energy for luminous figures on clock and watch dials. A fluorescent substance mixed with the radioactive element emits light when struck by a helium nucleus; and under high magnification it is possible to see the tiny individual flashes due to individual helium nuclei.

For forty years after the discovery of radioactive elements, however, they were regarded as a novelty, of great scientific interest, but with little practical importance. There was evidently a vast amount of energy locked up in the giant nuclei, but the naturally occurring isotopes decayed so slowly that 'atomic energy' – or, more accurately, 'nuclear energy' – seemed an impossible dream. But in 1938, the situation changed dramatically. German scientists Otto Hahn and Fritz Strassmann discovered that a very overweight nucleus like uranium could be split into two more or less equal-sized chunks (barium and krypton nuclei) if it were given a gentle nudge by a slow-moving neutron. The new, smaller, nuclei repel each with enormous force, again due to electrical repulsion, and as they shoot apart they collide with other nuclei and turn this energy into heat (vibration of atoms). Fortunately for the Allies, the Nazi government did not at first realize the implications of the research going on in the Kaiser Wilhelm Institute; and the information travelled along the scientific grapevine to the United States.

Here the recently arrived Italian physicist Enrico Fermi fitted the crucial piece into the jigsaw. (It was Fermi who discovered and named the neutrino, or 'little neutral one'.) Since the heavier nuclei contain relatively more neutrons, the split-up of a uranium nucleus might well release some of the excess neutrons. These particles could then nudge other uranium nuclei into splitting, releasing more neutrons, each with the potential for breaking up another nucleus. The resulting 'chain reaction' could produce a continuous supply of energy, if the number and speed of the neutrons were properly controlled – or an instantaneous cataclysm, if they were not.

The crucial requirement was that each fission of a uranium nucleus should produce at least one free neutron. Leo Szilard (who had had to emigrate from Nazi Germany) was one of the physicists who looked for evidence of neutron release, using apparatus in which the neutrons would produce tiny flashes on a screen. In his own words: 'We turned the switch, and we saw the flashes. We watched them for a little while and then we switched everything off and went home. That night there was little doubt in my mind that the world was headed for grief.'

The rest is history. Enrico Fermi built the world's first nuclear reactor in a disused squash court at the University of Chicago; at 3:45 pm on 2 December 1942 he withdrew the neutron-absorbing control rods and the chain reaction became self-sustaining: the nuclear age had begun. A nuclear bomb required very active fissionable material, so the Americans began large-scale extraction of the less stable isotope uranium-235 from the natural mixture, and they also started to produce large amounts of an artificial radio-active element, plutonium (which has two more protons than uranium). Wartime pressures led to the deployment of

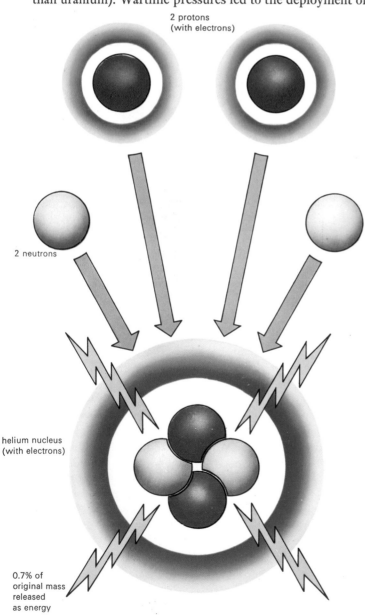

2 protons
(with electrons)

2 neutrons

helium nucleus
(with electrons)

0.7% of
original mass
released
as energy

half a million people on the project, and in July 1945, a prototype bomb was detonated near Alamogordo, New Mexico. Within a month, nuclear energy was unleashed in anger, as nuclear fireballs appeared above the Japanese cities of Hiroshima and Nagasaki.

Fitting it together: the states of matter
The main actors on the stage of the Universe have now introduced themselves: the simple electrons; the sly neutrinos; and the complex particles of the nuclei, each with its internal battle between strong and electromagnetic forces, and each hiding away its basic make-up, the mysterious quarks. At moments of high drama, a plethora of other particles – strange particles, charmed particles, and the whole sinister gang of antiparticles – make a brief appearance from the wings.

The long drawn-out plot, sometimes orderly, sometimes apparently in chaos, follows strict rules: relationships between the actors are governed by four forces, and the laws of Nature must not be overstepped. As we study the great drama of the Universe in more detail, we find that the four forces seem to be expressions of just one basic interaction between the characters: hate is love, if we have enough perception to understand it.

To complete the analogy, each of us is a subplot. In our bodies a particular combination of particle-players has been assembled to work out a tiny corner of the story of the Universe, before dispersing and regrouping in the action of another subplot. Man's life may have seven ages, but Nature's has a countless number.

Our own corner of the Universe is rather unusual, because it is extremely well ordered. Electrons revolve in neat orbits around nuclei to make atoms, and the atoms join together in an orderly manner to produce molecules. The regular stacking of sodium and chlorine atoms to make the perfect cubic crystals of ordinary salt seems almost a miracle: each crystal consists of 100 million million million atoms, and each fits exactly into place. Rocks, too, contain crystals, often brightly coloured; and even viruses – like the infectious flu 'germ' – have a beautiful crystalline form, on a microscopic scale.

The impressive orderliness around us springs from the fact that electron orbits are strictly controlled. Unlike a satellite circling the Earth, which we can put into any desired orbit, the quantum theory shows that on the very small scale of atoms, the electron can only whirl around the nucleus with whole numbers of 'spin units'. Hence only certain orbits are allowed. Filling up the allowed orbits with the right number of electrons for any element leaves the electron arrangement different from that of any other element, and so the chemical properties, hinging on the outermost electrons, vary from element to element.

*The site of the city of Nagasaki
two months after it was bombed.
The buildings still standing
in the foreground and background*

*mark the edges of the area of
total destruction: the flat
region between them was the
city centre.*

Look at common salt again. The chemist tells us that a sodium atom has a loosely attached outer electron, which the chlorine atom is keen to take over to complete a particular set of orbiting electrons. This transfer complete – in an explosive flash, if sodium metal is dropped into chlorine gas – the sodium has lost one negative charge, and is now electrically positive, while the chlorine atom is now negative. Such charged atoms are called *ions*. The two cling together as the opposite charges attract, and ultimately they fit together as a huge crystal *lattice*.

Atoms can also join together by sharing electrons. In chlorine gas, the atoms are so desperate to complete their electron sets that they band together in pairs, shuttling their outer electrons back and forth, so that each gets the impression of a complete set. Other gas atoms do likewise: the oxygen and nitrogen in the air are not individual atoms but molecular pairs (O_2 and N_2). Electron sharing happens in solids, too, and carbon is the most notable example here. Living matter is made of long strings and loops of carbon atoms, with occasional oxygen, nitrogen and hydrogen members, all bound together by the sharing of electrons. From the interplay of the electromagnetic force between shared electrons, the tied electrons of atoms, and the posi-

tive pull of the atomic nuclei, molecules acquire a definite shape. From these beginnings we end up with the beauty and complexity of living organisms.

The matter we know is traditionally classified into three states. In *solids*, the forces between molecules (or between individual charged atoms, in crystals) are so strong that they always retain their shapes, whereas the molecules in *liquids* attract each other enough to stay together, but can flow past one another. Molecules in *gases* are completely independent of each other, and fly about unimpeded.

We all know that states of matter are changed by heat. To take the most familiar case, we can warm solid ice and it will melt into liquid water; heat water and it will boil away as steam – a gas. Each time we add heat energy the molecules jostle each other more violently as they vibrate more forcibly. First the orderliness of a solid is lost at melting; and then at boiling the molecules are moving fast enough to lose contact with their neighbours, and speed off as free gas molecules.

So all matter on Earth is an orderly arrangement of atoms in molecules, whether the molecules are tied down, or freely moving in a gas. But Earth is only a minute part of the Universe – and it is something of a haven, for most of the

Three states of matter: ice and water are the solid and liquid states of the same substance. Air is a gas, composed of

different molecules – nitrogen and oxygen. The fourth state of matter, plasma, does not exist naturally on Earth.

matter in the Universe is in infernally hot regions, where the temperatures range from thousands of degrees to hundreds of millions of degrees. Under these harsh conditions, the jostling of atoms and molecules becomes a series of tremendous collisions; molecules are smashed up into atoms. And even atoms cannot survive as such. The collisions strip off their electrons, and leave the bare nuclei and free electrons to pursue their independent paths.

This type of matter – freely moving nuclei and electrons – is called a *plasma*, and most of the matter in the Universe is in this state; it's sometimes called 'the fourth state of matter'. Scientists can produce small amounts of plasma in the laboratory, but high-temperature plasma is a tricky substance to deal with. It can only be controlled by a magnetic field, and no sooner is the plasma concentrated than the field becomes unstable, and whips the intensely hot substance about inside the apparatus.

Out in the Universe, matter can take even more bizarre forms, if conditions are right. Squash the plasma making up a star sufficiently, and it can reach a state where the pressure is supported by the packing together of electrons; squeeze even more, and the electrons combine with protons to leave an even tinier neutron star. Imagine in the first case a star like the Sun compressed to the size of the Earth; and then, in the second, to the size of Malta, or Brooklyn. These are ideas which modern astronomers and physicists have to feel at home with; and we shall look at these oddities later, for they are the remnants of star death.

Saturn, with its wide set of
rings, is the most beautiful
planet in the solar system. The
rings are composed of millions of
rock and ice fragments, all
circling the planet as independent
satellites. Copyright by the
California Institute of Technology
and the Carnegie Institution of
Washington. Reproduced by
permission of the Hale
Observatories.

4:The Other Planets

'That's one small step for a man, one giant leap for Mankind.' On 21 July 1969 Neil Armstrong became the first human being to set foot on a world outside his own: after three thousand million years sequestered on Earth, life as we know it has begun to spread its interplanetary wings. Perhaps this remarkable achievement will eventually lead to human colonization of other stars' planetary systems, or even to contact with other celestial civilizations? At present this is just exciting speculation. But the manned Apollo Moon flights did bring back a huge amount of scientific information, which is helping astronomers to piece together the early history of the solar system.

Our planetary system consists of nine major planets, and vast quantities of smaller boulders, rocks, grains and dust – the 'rubbish' of the solar system – all orbiting our local star, the Sun. Before looking at each planet in detail, let's follow the solar system through its history. Starting with its formation from a huge dust and gas cloud, we can trace the planets' stories and discover why the planets have ended up so different from one another today – giant gas-and-liquid Jupiter; barren, scorched Mercury; hot, poisonous Venus; or, nearest to Paradise, planet Earth.

Familiar Earth, the cradle of Mankind, is naturally the best studied planet, and it might seem the obvious place to begin looking for clues to the origin of the planets. But despite the dedicated labours of geologists and other 'Earth scientists', our planet conceals its past uncannily well. The seemingly permanent rocks around us are all quite recent in comparison to the Earth's great age, many being barely one-tenth as old. For the Earth's surface is constantly changing. The rate is generally too slow for us to notice in the short space of a human life, but it is dramatically fast to the geologist who is used to dealing with spans of a thousand million years. The Earth's continents and ocean floors are drifting about, and thrusting up mountain chains when they collide; hot inner rocks break out from volcanoes, and push up at 'seams' in the ocean beds; the ocean-floor rocks are absorbed back into the Earth at the deep ocean troughs. And existing land is continuously being weathered away by the persistent scraping of running water, the eroded rock being washed back to the ocean. Here it will lie until another paroxysm pushes it up as a new piece of 'solid' ground. Small wonder then that geologists can say little about the Earth's birth or its infancy.

We must seek clues to the beginning of the Earth and the other planets elsewhere. There's no doubt that all the planets formed at much the same time, and we can piece together the evidence that each offers us. The Moon seems a good starting place. Quite apart from being the only other body which manned expeditions have reached, it is a superb fossilized museum of the early stages of a planet's life. 'Planetologists' regard the Moon as a planet, for it is not out of place in the league of rocky planets near the Sun. Fully one-quarter as large as the Earth in diameter, it is very much larger in comparison to its controlling planet than any other satellite; and an imaginary astronomer on Mars would see the Earth and Moon as a rather unequal 'double planet'.

The Moon's surface does not suffer constant ploughing by drifting continents, nor does it have running water, or indeed air, which could erode away its surface rocks. Scientists expected that the rocks brought back by the Apollo astronauts – over one-third of a ton in all – would be very old indeed. But lunar rocks are only as ancient as the oldest Earth rocks. The Moon has hardly any young rocks, as expected, but it also has very few extremely old ones. The reason turns out to be that the lunar surface suffered continuous battering by a rain of rocks from space for hundreds of millions of years after its birth, scarring it with the craters that we see now, and obliterating its original surface.

And so it turns out that planetologists have reason to be grateful for the rocky rubbish which litters our planetary system. The Earth acts as a gigantic dustbin, collecting rock fragments at a rate of a hundred tons a day. Small bits the size of sand grains burn up as they fall through the air, to shine out briefly as 'shooting stars' or *meteors*, while the rarer larger chunks can light up the whole sky as a fireball. Although most of a fireball is boiled away by its fiery passage through the atmosphere, a small remnant may survive to strike the Earth as a *meteorite* fall.

One rare type of meteorite is the solar system's 'birth certificate'. These dirty-looking, crumbling rocks – going by the tongue-twisting name of *carbonaceous chondrites* – are the oldest rocks known anywhere: unaltered since the birth of the solar system, they allow scientists to place that momentous event 4,600 million years in the past.

Dating the rocks

It's not obvious at first sight how anyone can tell the age of a rock – or even to decide what you mean by its 'age'. Pick up a stone in the garden, and it has no obvious clues. A sedimentary rock, like limestone, may contain the fossils of creatures and plants which died and settled into ooze on the sea bed, which was later compressed to rock. Identifying fossils is the 'classical' dating method in geology, since particular species lived only at specific times in the past. But where fossils are absent, as in volcanic and very old rocks – not to mention lunar samples and meteorites – dating must rely on more subtle tests.

Radioactivity holds the key. Since the unstable nuclei of atoms such as uranium, which are present in small amounts in the rock, are continuously breaking up into more stable ones, such as lead, at a rate which nuclear physicists have

*Astronaut James Irwin with the
Apollo 15 Lunar Rover. This
vehicle was driven a total of
28 km (17 miles) on the Moon,*

*and the astronauts collected
76 kg (168 lb) of surface rocks
and soil from the Hadley Rille
region.*

measured, the principle of radioactive dating seems very simple. Measure the present levels of uranium and lead in the rock, and calculate how many years must have passed to allow that proportion of uranium to decay.

Putting the principle into practice, however, is not quite that simple. There's the problem of deciding how much of the lead in the rock was there originally, and not formed by the decay of uranium. And during the history of the rock sample, some of the lead atoms may have migrated from their original sites. Fortunately, Nature has given us some help in dealing with these problems; for there are two distinct isotopes of uranium, which decay into different lead isotopes. Uranium-238 loses eight helium nuclei to turn into lead-206, while the shorter-lived uranium-235 – the active material of the first 'atom' bomb – sheds only seven helium nuclei to end up as lead-207. (The rarity of this second uranium isotope, incidentally, is entirely due to its shorter average life: during the existence of the Earth, some 99% of the original uranium-235 has changed to lead, while

only half the uranium-238 has decayed.) Both types of uranium will have undergone the same chemical reactions, and so the planetologist has two different 'clocks' to compare, both 'set' at the same time, and running at different, but known, rates. Ironically enough, the main problem now facing dating laboratories in university cities is contamination by airborne lead from car exhaust.

The uranium-lead age for the oldest meteorites is some 4,600 million years. This dating tells us when the meteorite's material acquired its present form, because any major mixing-up or complete melting would redistribute the uranium and lead, and 'reset' the clocks. For this reason, all other rocks in the solar system turn out to be younger. The Moon's surface dates back to a period 3,200 to 4,200 million years ago, long after the Moon first formed, but a time when falling rocks blasted out its craters and pulverized its surface. The Earth's oldest surviving rocks that we know of are found in western Greenland, and were formed about 3,800 million years ago.

The first chapter in the history of the solar system – the first few hundred million years – is missing from the planets' rocky records, and only the dirty, frail carbonaceous meteorites can provide us with real clues.

A plethora of clocks

The uranium-lead clocks are not the only radioactive clues of use to the planetologist: any radioactive atom with average life of several thousand million years will do. Thorium decays to lead, and rubidium to strontium over this time scale, and these clocks provide a valuable check on rock ages.

The other important radioactive clock is a rare isotope of potassium, whose nucleus can decay by the unusual expedient of catching one of its orbiting electrons, and thereby converting one of its protons into a neutron (just the opposite process to the decay of a 'free' neutron, in fact; but, in terms of energy states, favoured in this nucleus). The result is argon-40. Argon is a gas, the third most common constituent of the Earth's atmosphere after nitrogen and oxygen, and the 1% of argon in the air we breathe has come from the decay of potassium-40 atoms in the Earth's crust over geological time. It has an important practical use, for argon is *inert*; that is, it is extremely reluctant to take part in chemical reactions and so is used as filling for electric light bulbs.

In recent years planetologists have been able to push right back into the period before the solar system formed, by an extension of the radioactive clock method. All the heavy nuclei, including the radioactive ones, were made inside stars and thrown out into space by stupendous supernova explosions. The individual nuclei which break up in radioactive decay are thus considerably older than the solar system. This is irrelevant to traditional radioactive dating methods, which compare the relative numbers of radioactive and 'daughter' nuclei, produced since the parent atoms were trapped in the rock. But supernovae also jettison unstable nuclei into space: isotopes with an average life of about a million years. If the solar system formed soon after a nearby supernova explosion, these nuclei would have been included in early meteorites, and their decay products, the daughter nuclei, should still be locked up, even though the original unstable nuclei have long since decayed.

The first analyses suggested that there was a gap of some hundred million years between a supernova stirring up the pre-solar system gas cloud, and the birth of our planetary system. But in the past few years, Gerald Wasserburg and Typhoon Lee of the California Institute of Technology have discovered an unexpectedly large amount of magnesium-26, the daughter product of unstable aluminium-26, in an ancient carbonaceous meteorite which fell near the Mexican town of Allende in 1969. The original aluminium-

26 must have come from a supernova which exploded only a few million years before the meteorite formed. And this is such a short time scale in astronomical terms that the supernova explosion and the birth of the solar system may well be linked.

Other solar systems

This is a natural point at which to stand back and look at the formation of the solar system from a more cosmic viewpoint. Since our Sun is a very typical star, we might expect that other stars also have planetary retinues; by looking at the dense clouds of gas where stars form, we may learn something about the general way in which planets are born. When scientists eventually work out the full story of how the Earth and the rest of the Sun's family came into being, success will come by combining the story told by ancient meteorites with the discoveries made by astronomers investigating the birth of other planetary systems far off in the depths of our Galaxy.

For many years, astronomers thought that the birth of the solar system was a freak event – perhaps caused by the one-in-a-million close approach of another star to the Sun. Opinion has now swung right around, and according to modern theory it is only the odd star which is formed without planets. Could we settle the question directly, by looking to see if nearby stars have companion planets? Unfortunately, this is beyond the reach of present-day telescopes. Because planets have no light of their own, the members of another planetary system would shine only by light reflected from the central star, and the dazzle from the star would overwhelm the faintly glowing planets.

But there is an indirect way to tackle this problem. Peter van de Kamp, of the Sproul Observatory, Pennsylvania, has devoted his life to studying the slow movement of stars across the sky; and he has convinced many astronomers that some nearby stars do have planetary systems. A nearby, single star should follow a straight line in its path across the sky – very slowly. But if the star has an accompanying planet, in reality, both star and planet are orbiting their 'centre of gravity', the balance point of the system. The star moves in a far smaller orbit than the much lighter planet, but Professor van de Kamp can measure the slow, minutely small, swinging motion of the star from side to side of its average straight path – an unmistakable sign that the star has an unseen orbiting companion. From the size of the swing he can deduce the unseen planet's mass. And quite a high proportion of nearby stars turn out to have at least one planet a few times heavier than Jupiter.

Peter van de Kamp has discovered that the second nearest star to the Sun (after the triple system Alpha Centauri) has two planets of about Jupiter's weight. Since these could be taken as corresponding roughly to Jupiter and Saturn in our

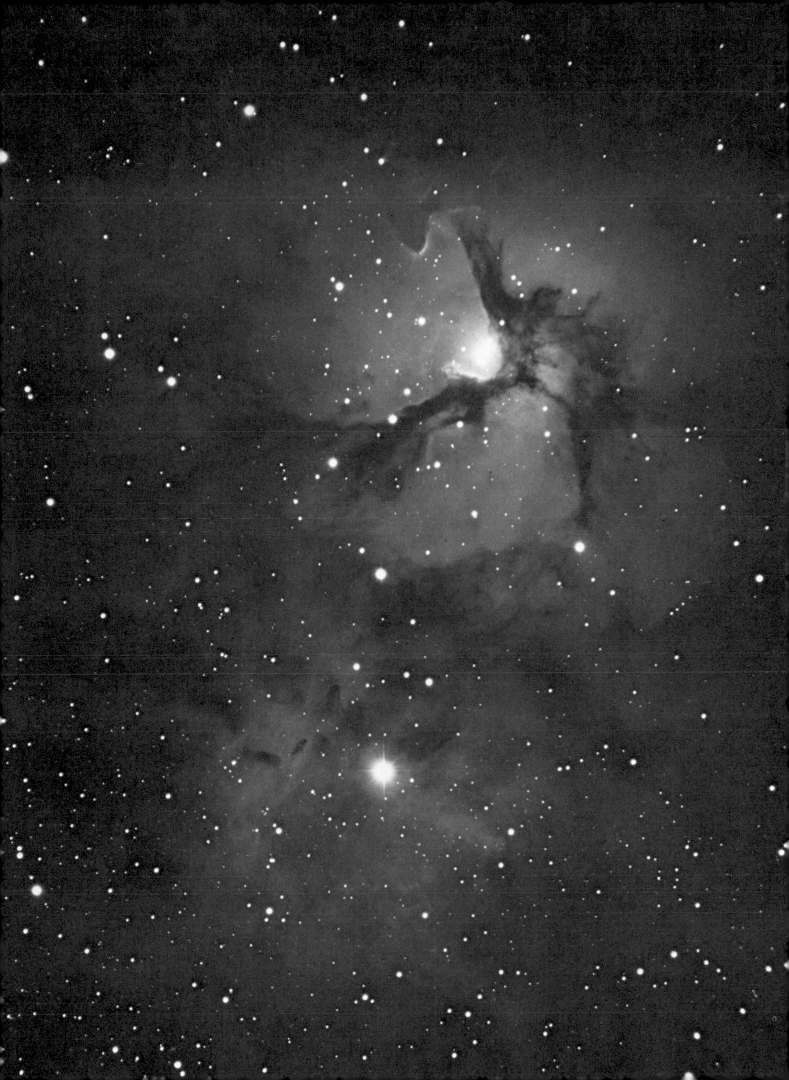

own system, it's tempting to think that Barnard's star may also have a troupe of smaller planets. This star is named after one of the most keen-sighted of modern astronomers, E. E. Barnard, and it holds the record for speed across the sky (relative to Earth), taking only 180 years to travel the apparent diameter of the Moon.

Planetary systems seem to be common, then, and this is quite in line with modern theories of star formation. These predict that a lot of grains of solid 'dust' will be left in orbit around a forming star, along with clouds of gas. As recently as 1977, astronomers studying the infra-red (heat) radiation from a star-forming nebula in the constellation Cygnus found evidence for just such a disc of dust and gas around a newly forming star. The planets should thus form quite naturally from this orbiting raw material.

To complete this quick survey of the solar system's pre-history, we must anticipate a little of the story of star formation, which we'll follow in more detail in Chapter 8. Large gas clouds are quite stable in our Galaxy, for the pressure of gas atoms whizzing about inside them is quite enough to prevent the gravitational pull between atoms from drawing the cloud together more and more tightly. A cloud like this will only begin to collapse and form stars if it is suddenly squeezed – and one of the most efficient ways of doing this is to hit the cloud with the high-speed gases thrown out by a supernova explosion.

So the ends begin to tie together. About one-third the lifetime of the Galaxy ago – 4,600 million years to be precise – a large cloud of interstellar gas was squeezed, and began to collapse under the inpull of its own gravitation. Just as astronomers see in other gas clouds which are collapsing now, this raw material then split up into several hundred stars, each surrounded by a spinning disc of dust particles and gas atoms. One of these stars was the Sun, and from its disc the Earth and the other planets were formed. The initial squeeze could well have been the shock wave from a nearby exploding star – both star-formation theory and the exciting new results on aluminium-26 in meteorites suggest that the birth of our solar system was the legacy of a nearby suicidal star which destroyed itself in a colossal supernova explosion.

The beginnings

At this point in the saga, events are still not entirely clear. Astronomers agree that the early Sun must have been surrounded by a rotating disc of dust particles and gas, and that *somehow* planets formed from the matter in the disc. But the exact details are in dispute.

The story may well have gone something like this – the version championed by A. G. W. Cameron of Harvard University. The tiny dust grains in the rotating disc were just those dust particles which are found throughout the space between the stars, and which block off the light from distant stars. But dust grains and atoms of gas are not the only occupants of interstellar space. Radio astronomers have found that the dense gas and dust clouds where stars form are rich in different kinds of molecules, mixed in with the gas atoms. These molecules can be readily identified because each type 'broadcasts' – gives off radiation – at a particular radio wavelength, like different cosmic radio stations to which the radio astronomer can tune in.

The molecules probably coat each tiny dust grain with a thin, sticky layer – indeed it's likely that the gas atoms react together to form molecules on the grain surfaces. As they were brought close together in the pre-solar-system disc, the grains collided and stuck together, to build up into loose bundles of dust, up to a centimetre (less than half an inch) across, as light and open as thistledown. At this time the Sun itself was still forming from the gas falling into the centre of the disc, and it had not begun to shine: the solar system was a very dark gas cloud, filled with countless millions of dirty thistledown balls wheeling around its centre.

Looked at from the fundamental viewpoint of particles and forces, the solar system's history is quite a simple process. In the original dust grains and their sticky jackets, we are starting with atoms joined together as molecules and larger crystals; and the accumulation of dust into planets just means piling more and more atoms together.

We've already hinted that gravitation is the prime force drawing the original gas cloud together. Random swirling motions in the original huge nebula, which broke up into hundreds of new stars, left each of its smaller collapsing fragments – including the one which became our Sun and solar system – rotating slightly. As gravitation drew it together into a smaller and smaller volume, the speed of rotation increased, to keep the total amount of 'spin' in the system constant. It was for this reason that the cloud could not all shrink to the centre to become part of the forming Sun, but some was left as a surrounding rotating disc.

As the dust grains collided and stuck together, gravitation took a back seat, because this sticking is an interaction between the electron clouds of atoms brought into contact, just as the surface atoms of adhesive tape stick it to a sheet of paper. Here, the electromagnetic force takes on one of its many guises, and plays a vital role in making the planets.

But once we have reached the dirty thistledown stage, gravity once again takes over the role of master builder. A rotating disc of particles this size cannot survive for long: calculations show that the gravitational force between the dusty balls, small though it is, will make them bunch together into chunks of roughly a kilometre ($\frac{5}{8}$ mile) in size.

These rocky chunks can, in turn, bunch together to make bigger and bigger bodies; but they are now large enough for

collisions to be destructive, too. Instead of holding together under their combined gravity, two colliding 'planetesimals', as they are called, may smash each other apart. The build-up of planets from planetesimals was a demolition race around the forming Sun, as the whirling rocky chunks collided, smashing or amalgamating in the process. The largest planetesimals were least likely to be broken up, and they eventually won the race, capturing most of the rest of the planetesimals to build up rocky planets. So were formed the innermost planets, ranging in size from the Moon to the Earth. This inner set is completed by the intermediate-sized Mercury and Venus, nearer to the Sun; and Mars, which circles around outside the Earth's orbit.

As we saw before, the Earth and Moon are really a double planet, and astronomers are still at a loss to explain how this double system – unique in the solar system – could have formed. Once it was thought that the Moon may have been part of the Earth, but the composition of the returned Moon rocks is significantly different from those of the

Earth's surface. It now seems that the Moon must have begun as a planet in its own right; but no one knows yet if it was born along with the Earth, or if it has moved in from another part of the inner solar system to be 'captured' by the Earth.

The planets of the outer solar system are very different: they are made almost entirely of substances that we know as gases, but compressed to the liquid state. This takes us right back to the very beginning of the solar system, for the sticky dust particles making up the pre-solar-system disc were surrounded by vast quantities of gas; indeed, there may well have been as much gas in the disc as there was at the centre condensing to form the Sun. But the disc gas was not all at the same temperature. Near the centre, where the Sun was forming, temperatures ran as high as 1500°C (2700°F), and even at the Earth's distance it was some 300°C (570°F) – hotter than a 'very hot' oven. In these torrid conditions, only high-boiling-point solids could remain condensed on the grain surfaces, and the icy jackets which the grains would originally have had were evaporated away. So the inner planets are 'rocky' in composition.

But at Jupiter's distance and beyond, the grains retained their jackets of ice and frozen ammonia, and the 'thistle-down' balls that formed were more like dirty snowflakes. And so the planetesimals out there would have been icy, too: enormous, dusty snowballs several kilometres, or miles, in diameter. Out there, where the temperature was low, the gas atoms in the nebula were moving more slowly, and the outer planets were able to capture this hydrogen and helium to add to their bulk. Giant Jupiter carried this to extremes, ending up some three hundred times heavier than the Earth, with hydrogen making up almost all this mass. The other 'gas giants', Saturn, Uranus and Neptune, did not have Jupiter's success in harvesting the surrounding gas, and they have ended up as smaller worlds – although each of them dwarfs the Earth into insignificance. (As we shall see later, the furthest planet known, Pluto, is an oddity: a tiny runt in the family of gas giants.)

The icy wastes of the solar system

Armed with some knowledge of the family history, let's now meet the individual planets of the solar system fraternity; and let's start with minds cleared of our natural 'Earth chauvinism'. An alien visitor to our planetary system would find Jupiter far and away the dominating planet: it weighs more than twice as much as all the other planets put together. (In fact, Jupiter could be the reason for an interplanetary visitor to call. From the distance of Barnard's star, say, an astronomer with instruments like ours would be able to detect the twelve-yearly wobble in the Sun's position as Jupiter goes around its orbit. He might suspect that if the Sun has a large planet, it may also

have smaller ones, too small to cause a measurable wobble of the Sun, but the right size to support life.)

Jupiter is so massive that some astronomers call it a 'failed star'. In its initial collapse from planetesimals and gas, Jupiter's centre heated up under pressure – like the air in a bicycle pump – but it never became hot enough to begin the nuclear fusion reactions which keep a star like the Sun shining. Instead of the millions of degrees necessary, Jupiter's centre is at a temperature of 'only' 30,000°C (54,000°F).

Despite its high central temperature, the outer layers of Jupiter are under the frigid conditions that we expect this far from the Sun. The atmosphere has frozen clouds of ammonia at −140°C (−220°F), and they would be colder if they were not warmed by heat escaping from the interior. There is as much heat coming from inside Jupiter as the planet receives from the Sun, and some astronomers believe that the high central temperature is kept up by a gradual shrinking of the planet. Unfortunately, it's difficult to test this idea, because the shrinkage needed is only one millimetre (0·39 inch) per year!

The interior of Jupiter is a strange place. As well as its high temperature, the enormous weight of the overlying layers squeezes the centre to enormous pressures – a hundred million times the pressure of the Earth's atmosphere. Although the bulk of Jupiter – apart possibly from a small rocky core – is made of hydrogen, captured as gas from the original nebula, this pressure squashes it into a totally unfamiliar form – liquid hydrogen *metal*. The ordinary liquid hydrogen which can be made at low temperatures and high pressures in a laboratory is just a tight packing of hydrogen molecules; but in Jupiter's harsh central conditions, such molecules are broken up into hydrogen atoms. In such a dense assemblage of atoms, some individual electrons can break free and conduct electric currents, just like a metal. Continuous churning of Jupiter's core produces great currents, and these generate a magnetic field in the same way as an ordinary electromagnet. The space probes Pioneers 10 and 11 measured this field, some ten times stronger than Earth's, when they flew past the planet in 1973 and 1974.

The most spectacular feat of the Pioneers was to send back extremely detailed colour pictures of Jupiter. These show the top of its atmosphere, for Jupiter has no real surface. Below an atmosphere a thousand kilometres (620 miles) thick and filled with clouds of all kinds, there is a gradual thickening of cloud layers until they merge into a deep 'sea' of liquid molecular hydrogen. Further down still, the pressure is high enough to make the liquid hydrogen metallic. Apart from a possible rocky core no larger than the Earth, Jupiter is a liquid planet, whose outer edge merges into its relatively thin atmospheric layer.

The Voyager space probes now on their way to Saturn, Uranus and Neptune should reveal the secrets of these more distant worlds. Their internal structure must be rather like Jupiter's, though not as hot or compressed at the centre. But each of these planets has its own odd quirks. Saturn, consisting mainly of hydrogen but not under Jupiter's vast internal compression, has the lowest density of any of the planets. Being less dense than water, Saturn would float if we could find a large enough ocean to drop it in. Uranus spins around an axis at right angles to its orbit, bowling about the Sun on its side, so that from the Earth we sometimes see the equator, and sometimes we are looking down on the poles. And Neptune, though much smaller than Jupiter, also seems to be generating some internal heat.

Snowball moons

We can't leave the giant planets without mentioning their weird and wonderful satellites. As befits its majesty, Jupiter has the most moons: thirteen definitely known, and a fourteenth suspected. Four of the satellites are large enough to be seen with binoculars, and some people can see them with the unaided eye in good conditions. Indeed, these four satellites would be easily visible if they were not overwhelmed by Jupiter's brilliant glare. Although they are tiny compared with Jupiter, the four large satellites are no dwarfs. All are larger than our Moon, and two are bigger than the planet Mercury.

Even though the satellites of the outer planets appear as little more than points of light to Earth-based telescopes, they have far more character and individuality than Earth's rocky satellite. Out in the depths of the solar system, ices as well as rocky dust particles agglomerated into satellites, and here the temperatures are low enough for some satellites to have kept an atmosphere. To see why this is so, let's think about the small planet Mercury, scorched by its proximity to the Sun. Mercury has lost any atmosphere it may once have had, for its high temperature means that the atoms of gas making up the atmosphere would have been very energetic: they would have moved so fast that they shot off into space, despite the planet's gravitational pull. Take Mercury to the outer reaches of the solar system, and even its weak gravity would suffice to hold down the sluggish atoms at these sub-zero temperatures, clothing the planet in an atmosphere.

Jupiter's third largest satellite, Io, is one of the most bizarre worlds in the solar system. About the size of our Moon, it is the smallest body known to have an atmosphere – albeit a very tenuous one, a thousand million times thinner than the Earth's. Its surface is reddish brown in colour, deepening to darker red at the poles, and it is covered with crystals of common salt, probably deposited when early oceans dried out. Io travels around an orbit within Jupiter's magnetic field, and so it suffers a continuous onslaught from protons and electrons – electrically charged particles from the Sun, which the magnetic field has captured as they stream past Jupiter. Theoretical astronomers had predicted that these particles should eject sodium atoms from Io's crystalline surface, to produce a huge sodium cloud around the satellite. We all know the yellow glow of sodium vapour in yellow street lamps: and Robert Brown of Harvard University searched for the yellow light of sodium near Io. In 1974 his quest was rewarded: Io is indeed surrounded by a cosmic-scale sodium lamp, a dimly glowing yellow cloud surrounding the reddish-coloured satellite. From Io's ochre 'salt-flat' surface, the night sky would glow yellow, while Jupiter would appear to hang in the sky as a huge striped oval. A science-fiction scene, but one which men may sometime witness; for although a manned landing on the giant liquid–gas outer planets is impossible, their satellites are natural staging posts for Man's exploration.

Contender for the title of strangest satellite is the largest member of Saturn's family of ten, Titan. A giant among satellites, intermediate between the planets Mercury and Mars in size, Titan has an atmosphere as thick as the Earth's, but composed of hydrogen and methane. Floating in this poisonous atmosphere are reddish-brown clouds, while pools of liquid ammonia may lie on the frozen surface. This picture of Titan is based on the most likely interpretation of observations made from Earth; but our knowledge of this world should be improved dramatically when Voyager 1 takes a close look as it speeds past Saturn in late 1980. It may also solve the riddle of Iapetus, a small satellite of Saturn which is as bright as snow on one side, and as dark as coal on the other.

Passing Uranus, with five smallish moons, we find that

one of Neptune's two is a large moon, similar in size to Titan. Similarly-named Triton may well be the largest satellite of all, but its distance keeps its secrets hidden from us. Strangely enough, Triton orbits Neptune the opposite way to all the large satellites of the other planets. And this may hold the key to another mystery: the tiny outermost planet, Pluto.

Although it is in the realm of the gas giants, Pluto is a tiny world, less than half the size of the Earth; indeed, recent indirect measurements suggest that it may be only the size of our Moon. And its orbit around the Sun is far from circular. For twenty years in its 248-year journey around the Sun, Pluto is within the orbit of Neptune; so between January 1979 and March 1999, for example, Neptune will be the outermost planet of the solar system.

Some astronomers link these facts together in the intriguing suggestion that Pluto may once have been merely a satellite of Neptune, and that both Pluto and Triton orbited the planet in the 'right' direction. But, according to this theory, the two satellites were in orbits which led to their gravitational pulls gradually disturbing each other, until a final catastrophic encounter hurled Triton about, to circle Neptune the other way, while the lighter Pluto was flung out into space to live a separate existence as a planet in its own right.

Shiny and tarnished rings

Although the outer planets are all similar globes, Saturn has one feature which sets it apart: an exquisite set of rings girdling its equator. In a small telescope Saturn is the most beautiful planet, and some would say the most magical

sight in the sky. To a Saturnian living above his planet's cloud layers, the rings would be a glorious permanent rainbow spanning the whole sky.

Despite their appearance in photographs and through the telescope, the rings are not a solid sheet. They consist of countless millions of tiny fragments of ice and rock – from about one centimetre up to one metre (half an inch to three feet) in size – each following its own orbit around the planet. Bright snowflakes and snowballs of water ice reflect light strongly, and make the rings shine: dark, larger rock chunks only show up when astronomers reflect radio waves from the rings and receive them as faint radar echoes. Although the total diameter of the rings measures 275,000 kilometres (170,500 miles) across, they are only a few kilometres thick: a piece of paper with the same proportion of thickness to width would have to be five metres (about sixteen feet) across.

Seen through a telescope, Saturn is the only ringed planet. But in 1977, airborne astronomers flying over the Indian Ocean had just that combination of luck and careful planning which has produced many a scientific breakthrough: they found that Uranus, too, has rings. The story began in 1973, when Gordon Taylor of the Royal Greenwich Observatory calculated that Uranus would pass in front of a faint star on 10 March 1977. Astronomers leaped at the chance to observe this rare event, for the way a star fades as a planet moves in front of it can reveal much about the planet's atmosphere. But, barely a month before the predicted date, refined measurements showed that the whole event would only be seen from the southern Indian Ocean, where there were no astronomical observatories.

Astronomers had an answer in the shape of the Kuiper Airborne Observatory, a high-flying aircraft equipped with a stabilized telescope, and named after one of the greatest twentieth century planetologists.

On the night of March 10, the Airborne Observatory took off from Perth, Australia, for a ten-hour flight at 12,500 metres (41,000 feet) above the desolate ocean. All eyes were fixed on the telescope, where the tiny disc of Uranus could be seen closing with the distant star. They had started observations an hour early, to allow for any slight errors in the prediction, and this careful planning bore unexpected fruit. Forty minutes before occultation (being hidden), the star suddenly 'winked' briefly. A few minutes later it did so again, and then again. Five times in all, invisible bodies had passed in front of the star and blocked off its light.

The planned experiment, to see the star disappearing behind Uranus, was duly carried out, but excitement centred around the unexpected disappearances. Would they repeat on the other side of the planet, when the star reappeared from behind Uranus? The Observatory stayed out an extra hour to see – and sure enough, the star faded again at exactly the same distances from the planetary disc.

Astronomers were due to watch the reappearance of the star from behind the planet in South Africa, and the airborne astronomers sent a message around the world – via Perth and Cambridge, Massachusetts – to advise the African astronomers to keep observing, and look out for the sudden winkings. When all the results were collected, the conclusion was inescapable: Uranus is surrounded by five or six rings, presumably of small orbiting fragments like those of Saturn. But there the resemblance ends. Saturn's rings are wide – the region around Saturn's equator is thick with fragments, and there are only a few clear gaps, which delineate the separate rings. Uranus' rings are narrow, however, each only some ten kilometres (six miles) wide (and probably about the same in thickness). Since they can't be seen through the telescope, even on very long-exposure photographs, they must be made of much darker fragments than Saturn's – there can be very few snowflakes and snowballs in Uranus' rings. If Saturn's rings were represented by a circular sheet of paper five metres (sixteen feet) across, the rings of Uranus on the same scale would be concentric circles of black cotton thread, the largest two metres (6·5 feet) in diameter.

Are these rings of rocky and icy fragments the remains of broken-up satellites which came too close to their parent planets? Or are they material which never assembled into satellites? Why are the rings of Saturn and Uranus so different? Does Neptune have faint rings too? The spacecraft Voyager 2 may help us to answer these questions, for if all goes well it will pass these three planets during the 1980s, sending back priceless information.

Comets: messengers from beyond the planets

Way out beyond the orbits of the known planets is the frozen realm of the comets. Most people's idea of a comet is a glorious spectacle of light, with a long tail appearing to stretch halfway across the sky. True, some comets do achieve this remarkable appearance – regarded as a sign of doom by our ancestors – as they sweep past the Sun; but it is really all show. The comet itself is a ball of ice and dust, typically the size of Malta – or Brooklyn. And most of its life is spent invisible out in the furthest depths of the solar system, in a path taking it perhaps halfway to the nearest star.

The origin of comets is still a puzzle. Their icy composition indicates a beginning in the outer parts of our planetary system; but were they thrown out from a birthplace near Jupiter, by that giant planet's gravity? Or are they remains of other, failed planetary systems which once began to form

near the Sun's family? Or perhaps they are actually made from gas way out in space, and are picked up as the Sun moves through the gas clouds lying between the stars. However they began, there must now be millions of comets pursuing enormously large orbits in the outer reaches of the solar system.

Occasionally a comet is perturbed – perhaps by the gravity of a passing star – and it takes an elongated path in towards the Sun. Picking up speed all the time, it heads in past the outer planets. Once within Jupiter's orbit, it is heated enough by the Sun for some of its ices to evaporate into gases and make up a huge, glowing head thousands of times larger than the original ice ball. Closer to the Sun still, the gas is swept away from the head by the continuous stream of protons and electrons from the Sun, to make a long, glowing tail. Meanwhile, the dust particles released by the evaporating ices are swept away by the pressure of sunlight to create a second tail. (Despite their impressive appearance, these tails are as tenuous as a good vacuum on Earth: in 1910 the Earth passed through the tail of Halley's comet without any effect at all.) Then the comet zooms past the Sun, and heads off back into the remote depths. As the tails must always point away from the Sun, the comet is then travelling tail-first.

Such a comet is moving in a very long orbit which won't bring it back to the Sun for thousands of years; but occasionally a comet's orbit is changed by a close approach to one of the giant planets. Swung into a smaller orbit by the planet's gravity, such a comet will circle the Sun in only a few, or a few dozen, years. And its fate is now sealed. On each swing around the Sun, some of the comet's mass is boiled away, and it gradually becomes smaller and feebler. Eventually it may break up altogether; or its centre may be left as a burnt-out 'clinker', endlessly circling around in the inner solar system.

The dust from a comet is a longer-lasting legacy, however. The small solid grains which are swept up by the Earth, and burn up in the atmosphere as meteors – 'shooting stars' – are dust from defunct comets. During its life, a comet strews out dust along its orbit: the much publicized but disappointingly faint comet Kohoutek of 1973/4 released over a thousand tons per second. When the Earth happens to intercept a comet's orbit, it naturally sweeps up copious amounts of dust, and a huge number of meteors can be seen. These meteor showers repeat on roughly the same dates each year – the beginning of August, for example, being a good time for meteor watching. Other showers vary in strength from year to year: the most spectacular ever recorded was on 17 November 1966 when, for a few minutes, observers in the United States were astounded by a veritable storm of meteors, falling at the rate of forty shooting stars every second.

Comets thus link the whole solar system. Coming in from the most distant parts known, they eventually die close in to the Sun, scattering their ashes to be swept up by the inner planets of our system – which brings us back to this inner, warm region of space near the Sun.

A failed planet: the asteroids

In surveying the early history of the solar system, we left the inner planets as the rocky planetesimals collided, shattered, and gradually agglomerated into five planets: Mercury, Venus, the Earth-Moon pair, and Mars. Outside the orbit of Mars, there's a region – the *asteroid belt* – still filled with fragments from the birth of the solar system. These asteroids, or minor planets, never managed to build up to planet size: indeed there is only enough matter in the asteroid belt to make a body one-tenth the diameter of the Earth. Half this mass is locked up in the largest asteroid, 1,000-kilometre (620-mile) Ceres, while the other asteroids range downwards in size from 500 kilometres. It's estimated that 50,000 asteroids could be picked up by present-day telescopes, and the two thousand brightest have precisely known orbits, and individual names. The first few were named after goddesses from classical mythology – Ceres, Pallas, Juno and Vesta – but these quickly ran out, and the list of asteroid names now includes those of countries,

The surface of the Moon is scarred by craters blasted out by infalling planetesimals in the last stages of its birth. The smooth plain is solidified lava,

which welled out subsequently. This picture was taken by Lunar Orbiter 3 in 1967, preparatory to the manned Apollo missions.

astronomers' wives, and even astronomers' surnames converted to a feminine form.

A few asteroids lie outside the main belt, and these have been distinguished by masculine names. Two groups which take the same orbit as Jupiter around the Sun are named after the heroes of the Trojan wars, the Greeks leading Jupiter and the Trojans following behind (except for Hector, who has somehow found himself in the Greek camp). Further out, circling the Sun between the orbits of Saturn and Uranus, is the asteroid Chiron, discovered as recently as 1977. Although it is far too small to be called a planet, its exact nature is an enigma: it could be simply a strayed asteroid, or the brightest of a new belt of outer-solar-system asteroids, or even the dead nucleus of a defunct comet.

Certainly many of the twenty or so 'asteroids' whose orbits bring them inside the Earth's orbit are believed to be dead comets. Among them is tiny Toro, which has been swung by Earth's gravity into an orbit taking it around the Sun exactly five times for each eight orbits of the Earth. (Because of this reference to Earth's gravity, sensational newspaper reports declared that astronomers had found the Earth's 'second Moon' – a complete misinterpretation of the discovery, in fact, for Toro circles the Sun rather than the Earth.)

While we are on the subject of moons, Mars' two tiny satellites are probably captured asteroids, and the close-up pictures of them, sent back by the brilliantly successful Viking space probes, showing them to be potato-shaped and cratered, give us a good idea of the appearance of typical asteroids.

Computers have been an invaluable aid to astronomers in the last twenty years, and nowhere more than in the monotonous business of calculating asteroid orbits from observations of their positions on the sky. Weeks of calculation can be reduced to a fraction of a second, and only in this way can astronomers keep track of the two thousand named asteroids. Computer analysis also shows that asteroids come in 'families' with basically similar orbits. It seems that the original dusty balls in the region between Mars and Jupiter accumulated to make up some fifty largish asteroids – or planetesimals, as we called them before – but none of them became large enough to draw them all together as a planet. Instead, collisions broke them down again into the myriad fragments that we see today.

But important changes had taken place in the meantime. In the centres of the planetesimals, the heat from the breaking-up of radioactive nuclei was trapped, and melted the original rocky material. Heavy liquid iron sank to the centre, while lighter rocks rose nearer the surface; and so when the asteroids later smashed apart, the fragments were very different. Some are pure metal, an iron–nickel mixture from a planetesimal's core; a few are metal–rock mixtures or rocky chunks from further out; but most common of all are fragments from near the planetesimal's surface, still in an unchanged state. Studies of sunlight reflected from asteroids show just the characteristics expected for such a mixture of fragments.

We can see the broken-up interiors of asteroids in many geological or natural history museums which have meteorite displays. These 'stones from space' are tiny pieces of asteroid which have strayed into the Earth's path, and have survived a brief trip through the Earth's atmosphere as a brilliant fireball. We can see 'irons', 'stony-irons', 'stones' and 'carbonaceous stones', coming from successively higher layers in the original planetesimals. It's difficult not to feel a sense of awe in the presence of rocks from outer space, the only extra-terrestrial material that most of us will ever see. And it's ironic that people stood for hours to see a tiny sprinkling of Moon dust brought back by the Apollo astronauts, while far larger, further travelled and older rocks from space are permanently on display in the meteorite sections of our museums.

The Moon and Mercury: scarred fossils

As planetesimals collided to form the 'successful' planets, the last to arrive smashed into worlds already almost formed. The colossal impacts caused vast explosions, which gouged out craters, circular holes with raised rims. The smallest planet, Mercury, and the Earth's Moon have both survived almost unchanged since this battering, and their surfaces are liberally pock-marked with ancient scars. We can see the Moon's largest craters, two to three hundred kilometres across, with binoculars; but these aren't its largest scars. The dark patches making up the face of the 'man in the Moon' are huge lava plains, many hundreds of kilometres across. They were gouged out by impacts of asteroid-sized blocks of rock about 4,000 million years ago. The dark lava surface we see today gradually oozed out to fill the huge holes as radioactive atoms melted the rock below the Moon's surface. It was this final battering by planetesimals, and the outflow of lava surface, which 'reset' the radioactive clocks which scientists use to date the Moon's surface, and make it appear younger than the 4,600-million-year-old meteorites.

The innermost planet, Mercury, looks remarkably similar to the Moon, according to photographs sent back by the Mariner 10 probe in 1974 and 1975. But Mercury is wrinkled, like an old apple. Long, winding ridges meandering across its lava plains show that Mercury has shrunk slightly since it formed. The reason is probably indirectly related to Mercury's closeness to the Sun: when it formed, the heat there was so intense that only high-boiling-point substances – in particular, iron – could condense. So Mercury has a large iron core, and as it cooled it shrank more than the rocky-cored Moon, wrinkling up the planet's surface.

Mariner 10's study of Mercury was literally a revelation. Until then, astronomers' glimpses of the planet's tiny disc, always near to the Sun in the sky, had shown only vague, dark markings – which were given mystic names, such as 'The Wilderness of Thrice-Greatest Hermes'. The markings were so difficult to see that astronomers were com-

pletely misled when they tried to work out the planet's rotation period. For a long time, it was believed that Mercury always kept the same face to the Sun, so that one hemisphere was perpetually baked, the other always frozen – an exciting scenario which generated many a science-fiction story. But when radar astronomers bounced radio waves off the planet in 1965, the returning 'echoes' showed that the planet rotates faster, exactly three times in two orbits around the Sun.

Because Mercury's orbit is not quite circular, a Mercurian would see the Sun follow a strange path across the sky. If he lived in the huge plain called the Caloris Basin, for example, he would see the Sun appear to rise in the east, move slowly across the sky for 33 Earth days, and then slow down when more or less due south. By the fortieth day it would have stopped, and it then would move backwards for eight days, before resuming its slow journey to set in the west after 88 Earth days. But our Mercurian would have to be a hardy creature. The 'noon' temperature reaches over 400°C (752°F), hot enough to melt lead, while the surface cools to liquid-air temperatures during the night.

Mars: abode of life?

The next smallest planet, Mars, has always seemed the best place for finding life outside the Earth: the word 'Martian' has become synonymous with invading extra-terrestrials. Through a telescope Mars certainly appears the most Earth-like of all the planets. It has red deserts and white polar caps, evidently of some kind of ice, which grow and contract with the seasons. Darker regions on the planet's face also vary slightly with the Martian seasons, and these were first thought to be seas, and then vegetation. Even though Mars approaches closer than any other planet apart from Venus, the details of these dark markings are elusive; to see them well requires a good telescope and keen eyesight.

This brings us to the story of Percival Lowell, an

American astronomer who devoted his observing life to Mars. Following an Italian report of thin, dark lines on Mars, Lowell founded an observatory to study the planet – and indeed, he could see hundreds of these narrow lines connecting Mars' larger dark areas. He was convinced that they were the work of intelligent Martians, and were large tracts of vegetation on the banks of canals which carried water from the polar caps to the arid deserts of the equator. The 'canals' were so straight that they must have been artificial; and since Mars seemed to be a drying-out world, what could be more natural than irrigation works on such a large scale? More imaginative minds began to fear that the Martians might leave their dying planet and migrate to our moist world. This was the inspiration behind H. G. Wells' story *The War of the Worlds* (1898) which, when it was adapted for radio by Orson Welles in 1938, was so convincing that many listeners thought it was a newscast.

But even in Lowell's day (he died in 1916) astronomers were sceptical. Even eagle-eyed E. E. Barnard simply could not see the canals, even with better telescopes than Lowell's. After Lowell's death, it became more and more apparent that Mars was much drier, and had less atmosphere than he thought; and the dark areas seemed less and less likely to be vegetation. Intelligent Martians seemed to be ruled out; but could simple life, like lichens, exist on Mars?

The first three space probes sent to Mars in the 1960s showed an even more barren planet than anyone had expected. Pictures sent back showed a Moon-like surface of craters – and certainly no straight canals. As it happened, however, all these early 'fly-by' craft had photographed the most uninteresting parts of the planet, and the situation changed when Mariner 9 went into orbit around Mars in 1971. Its startling pictures included an enormous system of canyons, 5,000 kilometres (3,100 miles) long, lying along the equator, with a main valley 75 kilometres (46 miles) wide and 6,000 metres (20,000 feet) deep. A comparison with Colorado's Grand Canyon is completely inadequate: this Martian valley would engulf the whole of the Alps, with room to spare.

Nearby are four giant volcanoes, dwarfing any volcano on Earth. The largest, Olympus Mons (Mount Olympus) is some 600 kilometres (372 miles) in diameter and 25,000 metres (82,000 feet) high – a single mountain which would cover the whole of Spain, and which rises to three times Mount Everest's height. Most interesting of all to those looking for life on Mars, Mariner 9 revealed many narrow, winding channels – undoubtedly dried-up river beds. Although the dark areas of Mars are not vegetation, but simply regions where the bright surface dust is blown away by seasonal winds, there was now a new reason for optimism in the search for life. The channels suggest that water once ran on Mars' surface; if so, the atmosphere must have

been denser then. Under these more hospitable conditions, simple life forms like bacteria might have developed, and these might now be lying dormant in the soil.

The two Viking spacecraft which landed on Mars in the summer of 1976 carried simple biological experiments to test for such life. The immediate outcome was widespread confusion, for the different tests gave contradictory results; but on weighing up the evidence it seems most likely that the 'positive' results were all due to chemical reactions with the unexpectedly reactive Martian soil. The deciding experiment was a sensitive test for compounds of carbon, the element on which life is based. None was found in the soil; and this result rules out living cells of the kind that we know on Earth. Most scientists have now reluctantly concluded that Mars – at least at the two Viking lander sites – is a dead world.

The hell-planet: Venus

Venus, although the Earth's nearest planetary neighbour, has always been a mysterious world. All that we can see through telescopes is the top of its continuous cloud cover, shining so brightly in the sunlight that it is the brightest object in our skies after the Sun and Moon. Venus is almost Earth's twin in size, and astronomers of the last century imagined that a paradise lay beneath its clouds. But we now know that Venus' surface resembles Hell; and planetologists say that our beautiful Earth would have gone the same way if it had been closer to the Sun.

While American space probes have made most progress in the study of Mars, it is Russian craft which have penetrated to the surface of Venus. The first attempted landers were crushed by a far thicker atmosphere than anyone had expected; but the later, heavily reinforced, landers succeeded. In 1975, two landers survived for an hour, and sent back the first pictures of Venus' surface. These Russian results, and those from the American fly-by probes, are disheartening to anyone thinking of travelling to Venus.

The beautiful clouds we see are probably made up of droplets of concentrated sulphuric acid. The atmosphere is a hundred times thicker than Earth's, giving an atmospheric pressure at Venus' surface the same as that pressing on a submarine 1,000 metres (3,280 feet) below one of Earth's oceans. The atmosphere is mainly unbreathable carbon dioxide gas, and the surface is at a constant temperature of nearly 500°C (932°F), even hotter than 'noon' on Mercury. The surface photographs show rock-strewn, barren plains stretching to the horizon, on a planet surprisingly light considering its massive cloud cover – it's been compared to an overcast day in Moscow.

The space-probe data have been complemented by radar results. Radio waves can penetrate the clouds; and radar 'echoes' showed, back in 1962, that the planet rotates very

The unbroken clouds of Venus show evidence for atmospheric circulation patterns when photographed in ultraviolet light. Mariner 10 took this close-up in 1974, as it headed past Venus on its way to Mercury.

slowly, and in the opposite direction to the other planets. If it weren't for the clouds, a Venusian would see the Sun rise in the West, and set in the East 60 Earth days later. More detailed analysis of the radar 'echoes' reveals enormous craters and mountains. The geological study of Venus is still in its infancy, and planetologists are looking forward to comparing the Earth with another planet of similar size.

But why is Venus so different from the Earth? The answer is that it is the Earth which is really the oddity: in its early days our planet was probably very like Venus' present state. So let's leave hostile Venus – a world where an intrepid astronaut would be simultaneously roasted, crushed, poisoned and corroded by acid – and head back to the only really friendly planet in the solar system, Earth.

Planet Earth

Our planet is third from the Sun, and the largest of the rocky planets. To a traveller from space, Earth would seem unique in other ways too. It's the only rocky planet without obvious craters, the only one with liquid water, the only one with plenty of oxygen in the atmosphere – and the only planet with life. All these oddities are in fact related. By

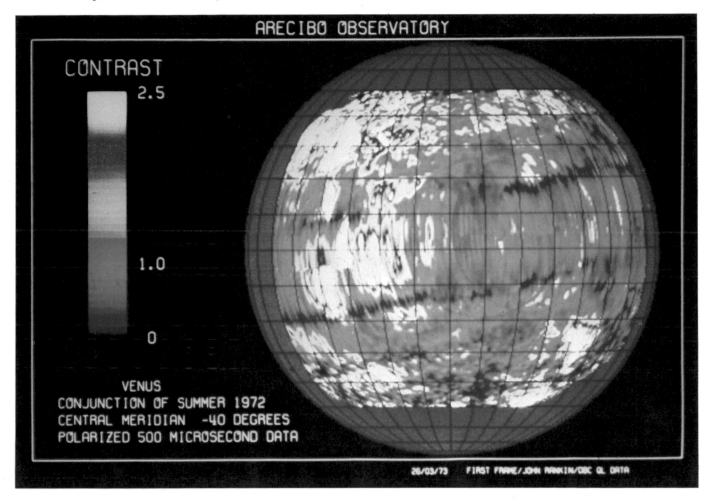

studying other planets, astronomers have come to the conclusion that the Earth owes its privileged status – and we our existence – to just two factors. One is that Earth's gravity is strong enough to hold down an atmosphere; and the other is its temperature, due to the Earth being just the right distance from the Sun.

An Earth too small to keep an atmosphere would have ended up much like our dreary companion-world, the Moon. Yet the Earth's virtual twin in size, Venus, has clearly taken its atmosphere to extremes – a choking gas blanket a hundred times denser than the Earth's. Why has our planet hit the happy medium? To answer this question, we must go back to the Earth's birth.

The early Earth must have suffered a final pounding from falling planetesimals right up to some 4,000 million years ago, just as the other inner planets did. In those days Earth must have been just a huge, scarred ball, like an overgrown Moon. And none of the planets could have had an atmosphere at this time, for as the Sun began to shine, it went through a period of superactivity. Protons and electrons from its surface blasted out through the early solar

system, sweeping away into space all the gas which was not incorporated into the giant planets, including any early atmosphere of the Earth's. But as the Sun settled down to its present relatively quiet state, volcanoes on the Earth began to surround the planet with a second atmosphere. Like the output of present volcanoes, these must have been unbreathable gases, mainly water vapour and carbon dioxide, but with sulphur compounds, carbon dioxide and hydrogen cyanide as unpleasant minor constituents.

Earth and Venus must have had similar atmospheres at this time, but the crucial effect of distance from the Sun soon became apparent. Earth was just cool enough for water vapour to condense into droplets of water, but Venus was too hot. On Venus the atmosphere just grew and grew, thick with carbon dioxide, and topped with clouds of sulphuric acid drops. The carbon dioxide acts like the glass windows of a greenhouse: it lets energy from the Sun shine in as light; but when this energy warms up the surface, the heat (infra-red) radiation cannot penetrate the gas to escape back into space. And so heat was trapped at the surface, and Venus' temperature rose swiftly to the melting point of lead. All Venus' water vapour has meanwhile escaped, probably broken up by sunlight at the top of the atmosphere, to leak off into space as individual oxygen and hydrogen atoms.

Earth escaped this fate. Water vapour condensed into clouds, which fell as rain. And carbon dioxide dissolves in water. As our atmosphere was built up by the steady out-gassing of volcanoes, most of it was able to dissolve in the seas; it could not keep on accumulating like Venus'. So the oceans saved the Earth from a heat catastrophe. They also altered the face of the Earth, for it is largely erosion by running water which has rubbed away the traces of Earth's ancient craters, and has left it the least pockmarked of the inner planets.

The other geological force shaping Earth's surface has been the slow movements of parts of its crust. Most school-children have noticed from their maps that the eastern 'bulge' of South America would fit neatly into the west coast of Africa; but only in the 1960s were geologists finally persuaded that it once had, and that continents really do drift over the surface of the Earth, breaking and reforming. According to the modern theory of continental drift, dubbed 'plate tectonics', the Earth's crust is made up of at least fifteen separate 'plates' (seven of them very large), all touching, and all gradually moving about. Most of the plates are just portions of ocean floor, but some of the larger ones carry the lighter continents on their backs. And as the plates move, their continental passengers drift across the Earth.

Where two touching plates are moving towards one another, something evidently has to give. In practice, one plate dives under the other and disappears into the hot interior of the Earth to melt. The grinding of one plate over another causes earthquakes above, while the newly melted rock can force its way up through the overlying plate to spew out as a string of volcanoes. The floor of the Pacific Ocean is a huge plate travelling north-west; and as it dives under the plate carrying Asia, it has thrown up a string of volcanoes which has grown into a row of earthquake-prone volcanic islands: we know this 'collision debris' as Japan.

The plates bearing the continents of South America and Africa are indeed drifting apart, remains of a larger 'super-continent', just as the school globe suggests. At the gap left in the middle of the Atlantic Ocean, new molten rock wells up from below and solidifies as a ridge which runs north-south along the entire length of the Ocean, breaking the surface as isolated islands such as Ascension Island and St. Paul Rocks. Further spreading drags this rock off to either side as part of the ocean floor, while newer rock wells up in the middle. The floor of the Atlantic Ocean is therefore youngest at the centre, becoming older as you take samples from further away on either side. It was this distinctive

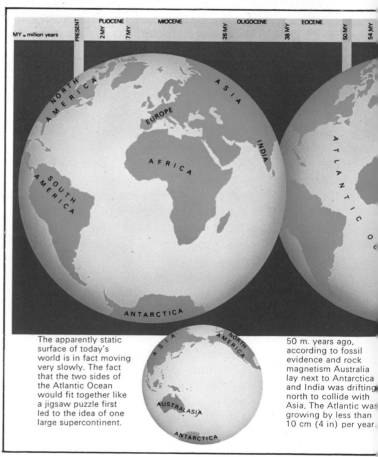

The apparently static surface of today's world is in fact moving very slowly. The fact that the two sides of the Atlantic Ocean would fit together like a jigsaw puzzle first led to the idea of one large supercontinent.

50 m. years ago, according to fossil evidence and rock magnetism Australia lay next to Antarctica and India was drifting north to collide with Asia. The Atlantic was growing by less than 10 cm (4 in) per year.

66

The Earth's appearance changes over millions of years as the continents drift over its surface. Geologists call the original supercontinent Pangaea with a large gulf The Tethys Sea. It split into two large land masses, Laurasia and Gondwanaland, which further fragmented into the continents we know today.

ageing pattern that first convinced many geologists that the plate tectonic theory should be taken seriously.

Another strong argument sprang from a study of the magnetism of continental rocks, impressed into them when they formed. If the positions of the Earth's magnetic poles are calculated from the magnetism of these old rocks, the results just do not agree. But if you move the continents and rotate them, a good fit can be made; and the positions of the continents at various times in the past tie in very well with the movements deduced from the ageing of the ocean floors. In the comparatively short space of ten years, geologists have made a complete about-face on the question of continental drift: from being a crack-pot idea, it has emerged within the wider theory of plate tectonics as the accepted way to explain the workings of the Earth's crust.

Since plates are continuously diving under other plates to be remelted, and new rock comes up at mid-ocean ridges, none of the ocean floors is more than a few hundred million years old – less than one-tenth of the Earth's age. The continents fare better, for they are composed of lighter rock which always stays on top. When two continent-bearing plates meet, they just pile up. The gradual forcing-together lifts up the edges into giant mountain chains, like

the Himalayas, raised when India crashed into Asia 'only' 50 million years ago.

There is some fascinating detective work involved in putting together the clues from ocean floors, the magnetism of continental rocks and the ages of old mountain chains – the mute witnesses of past collisions – to extract a picture of the Earth in the past. About 200 million years ago, all the continents formed just one huge supercontinent, which we call 'Pangaea' (Greek for 'whole earth'), and which subsequently broke up into the continents we know today. Pangaea itself was built up by a series of collisions between pre-existing continents during the preceding 200 million years. Geologists now think that plates have been on the move since at least 2,700 million years ago, and that their accompanying continents have travelled all over the Earth, sometimes as one supercontinent, at other times as individual rafts of lighter rock.

But what drives the moving plates of the Earth's crust? It certainly takes a lot of energy to move a whole continent at a speed of one to ten centimetres (one-half to four inches) per year, and expert opinion is divided. Ultimately, the source must be the heat released inside the Earth by radioactive atoms, but it is still not clear how this heat

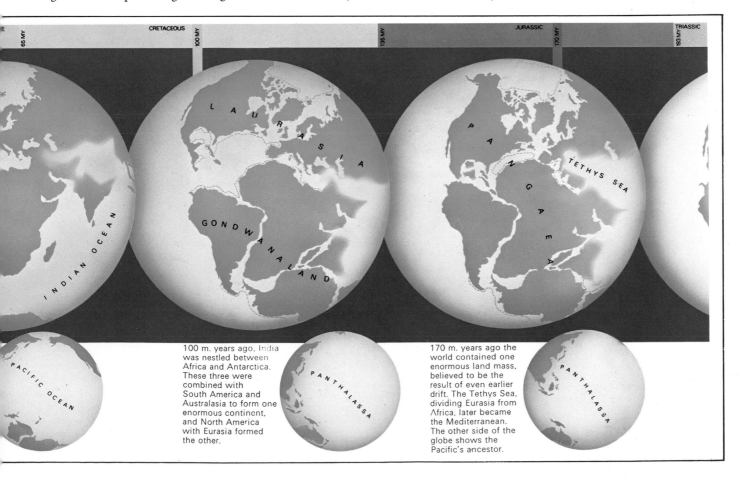

100 m. years ago, India was nestled between Africa and Antarctica. These three were combined with South America and Australasia to form one enormous continent, and North America with Eurasia formed the other.

170 m. years ago the world contained one enormous land mass, believed to be the result of even earlier drift. The Tethys Sea, dividing Eurasia from Africa, later became the Mediterranean. The other side of the globe shows the Pacific's ancestor.

*The Earth's drifting continents
are carried on larger moving
'plates'. The south Atlantic
Ocean has opened as plates
carrying South America and*

energy is turned into plate motion.

Within the Earth's crust, the heat released by uranium, thorium and potassium-40 causes the temperature to rise steadily. Underneath the crust, some 35 kilometers (22 miles) below a continent's surface, the rocks partially melt to form a 'slushy' layer. This layer lubricates the sliding of the crustal plates over the lower layers of the Earth, and makes the drifting considerably easier. Lower down still, the temperature is still rising, but the weight of overlying layers compresses the rock into a solid form. This zone, the 'mantle', makes up most of the Earth's interior, right down to a central core of nickel-iron. The outer parts of the metal core are liquid, and like the hydrogen-metal core of Jupiter, electric currents here generate the magnetic field which surrounds our planet.

Planetologists are intrigued by the Earth's massive geological activity. Does every planet of the Earth's mass automatically have a surface of shifting plates? When Venus is better studied, a comparison with Earth will help geologists to understand the Earth better. Already a comparison of Mar's volcanoes with the Earth's has told us something about

A glance at the map of the Atlantic coasts of Europe, Africa and the Americas re-veals a strikingly close match in their outline, as generations of school children have noticed, as though they were pieces of a jigsaw puzzle. In 1908 an Ameri-can geologist, Frank Taylor, published a theory that the continents had

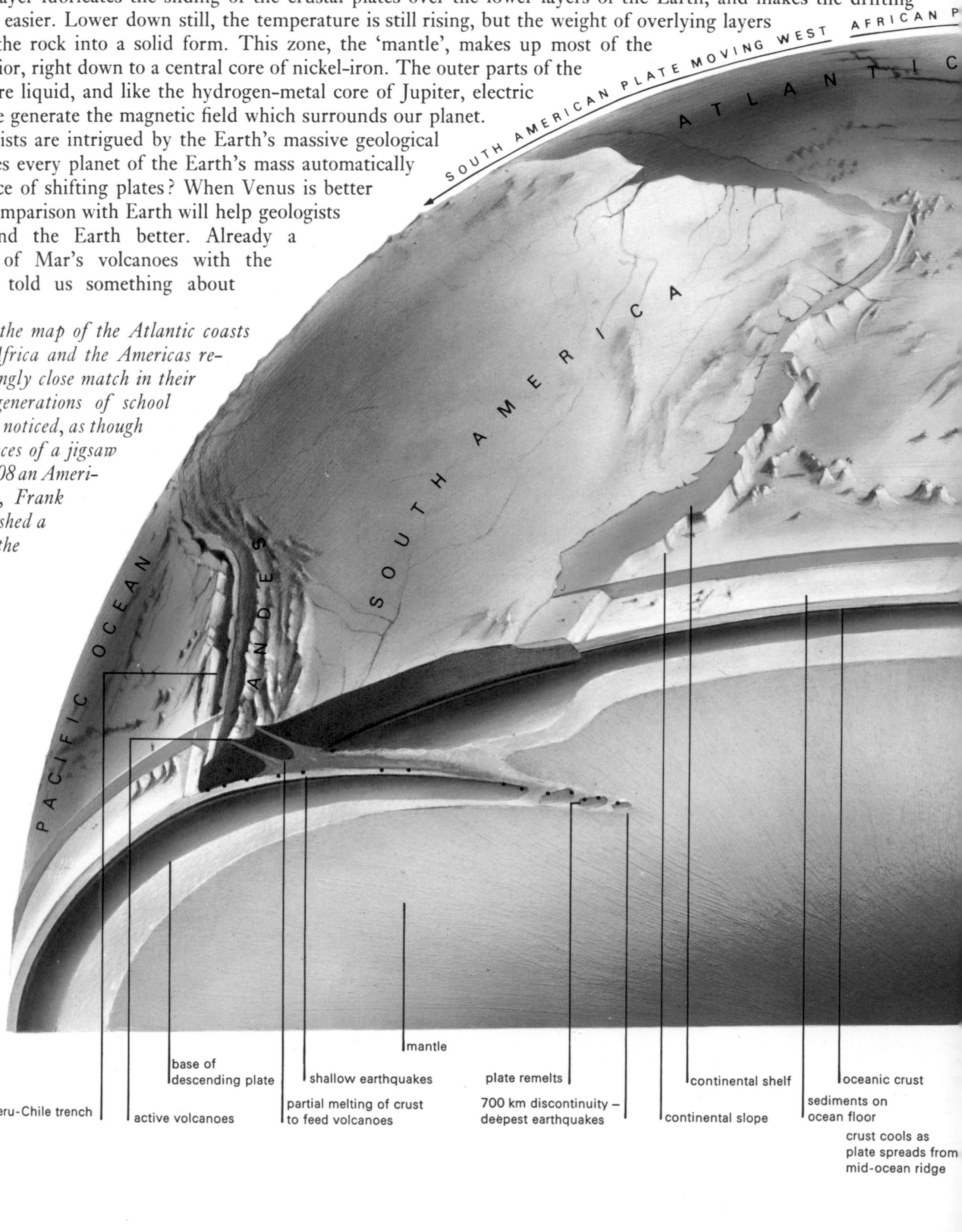

SOUTH AMERICAN PLATE MOVING WEST   AFRICAN P

ATLANTIC

PACIFIC OCEAN

ANDES

SOUTH AMERICA

Peru-Chile trench

base of
descending plate

active volcanoes

mantle

shallow earthquakes

partial melting of crust
to feed volcanoes

plate remelts

700 km discontinuity –
deepest earthquakes

continental shelf

continental slope

oceanic crust

sediments on
ocean floor

crust cools as
plate spreads from
mid-ocean ridge

Africa have drifted apart. New rocks well up at the mid-Atlantic ridge, while the Andes have risen as South America over-rides the Pacific plate.

'hot-spot' volcanoes.

Not all of Earth's volcanoes are caused by the Japanese situation of one plate sinking under another. Hawaii is a volcano in the centre of the Pacific plate, and it seems to be caused by a continuous upflow of hot rock from the mantle, always rising at the same place. As the Pacific plate has drifted north-west over this 'hot-spot', the rising column of rock has had to punch a succession of holes through the plate, while the old volcanic cones have been carried away in a long north-west line of islands and seamounts (underwater mountains not reaching the surface). Each volcano formed by the Hawaiian hot spot can only grow to a certain size before it is carried away from the rock supply. Mars shows no sign of plate tectonics, and the huge size of its volcanoes may simply be due to the continuing accumulation of lavas above a Hawaiian-type hot spot. Perhaps hot spots can develop on a wide variety of worlds, from Mars to ten-times-heavier Earth, but plate tectonics only on the biggest of the rocky planets. The study of Earth as a planet is a science still in its infancy, and there are no doubt many more surprises in store.

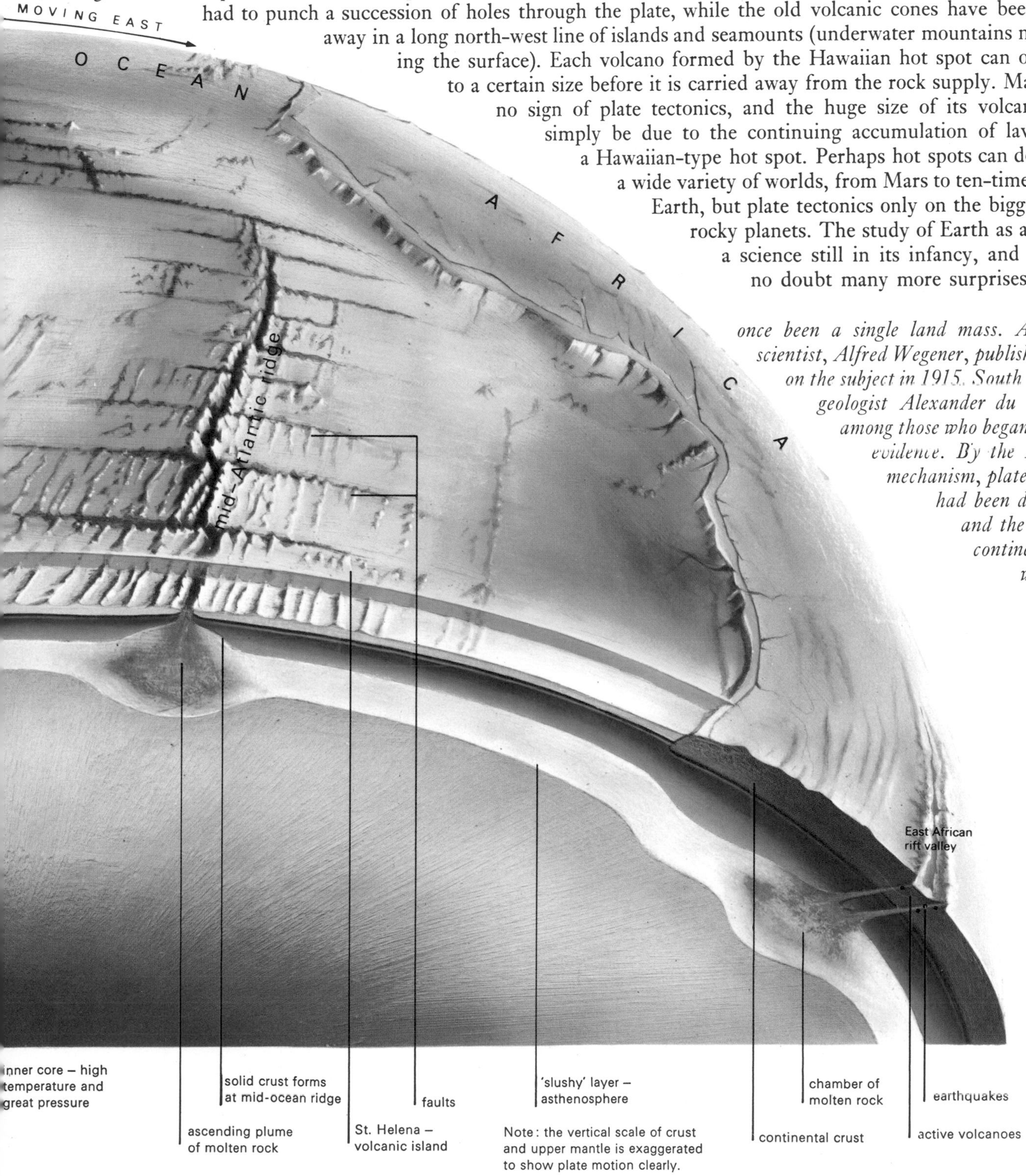

once been a single land mass. A German scientist, Alfred Wegener, published a book on the subject in 1915. South American geologist Alexander du Toit was among those who began to gather evidence. By the 1960s the mechanism, plate tectonics, had been discovered, and the theory of continental drift was widely accepted.

MOVING EAST

O C E A N

A F R I C A

mid-Atlantic ridge

East African rift valley

inner core – high temperature and great pressure

solid crust forms at mid-ocean ridge

faults

'slushy' layer – asthenosphere

chamber of molten rock

earthquakes

ascending plume of molten rock

St. Helena – volcanic island

Note: the vertical scale of crust and upper mantle is exaggerated to show plate motion clearly.

continental crust

active volcanoes

The complex arrangement of
atoms in the DNA molecules
(shown different colours in this
model) carry all the 'information'
needed to create and maintain
any living creature, from
microscopic plankton to Man.

5: Life: Organizing the Atoms

What's the difference between an elephant and a stalk of rhubarb? If we say 'very little', we might get a strange look from the grocer – yet this is precisely the answer which a modern scientist studying life would have to give.

Planet Earth is probably unique in the solar system in supporting life. And Earth's life forms are surprisingly similar in basic make-up, despite their huge variety of appearances. This conclusion has emerged after centuries of studies by biologists, and more recently by biochemists investigating the chemistry of life. These experiments have led to a tremendous amount of information about how terrestrial life forms are constructed, and how they work.

Modern astronomy has opened new doors to the study of life. First, astronomers and planetologists are rapidly building up a detailed picture of the Earth's formation, and can now say what conditions must have been like when the first living things appeared on Earth. Combined with the details of life chemistry now being elucidated by biochemists, the mystery of the origin of life on our planet is rapidly being solved. And the second new perspective is the realization that many other stars must have planets like Earth. So it's very unlikely that life is confined to our planet: many astronomers would say that our Galaxy must be teeming with life, and it is only the immense distances between the stars that keep us segregated from our brothers in space.

What would life from another planet look like? Would it be fundamentally like terrestrial life, or could it be based on chemical compounds and reactions totally different from Earthly life forms? Science fiction writers have certainly had a field day inventing alien creatures of all kinds, but we will start by looking closely at the kind of life that we do know: life on Earth.

Biologists, looking at living matter through the microscope, find that all living things – plant or animal – are made of tiny packages, or *cells*. A blade of grass is composed of millions of cells: the largest living creature, the blue whale, of millions of millions. The simplest organisms of all are just one single cell. Microscopic single-celled algae throng stagnant ponds, while the single cells of bacteria are well known to us as the 'germs' which are necessary to life, though some carry diseases. In the 'higher' organisms, cells have become specialized to perform different duties: long nerve cells carry electrical signals around our bodies; muscle cells have the ability to contract on demand; and white blood cells float around our bodies to catch and destroy invading bacteria. Yet they are all basically built in the same way. The human body can be compared to a large city: each worker – cell or citizen – is fundamentally similar, yet each has its own role to play in keeping the organism, or city, running efficiently.

Once we look inside the cells, the similarity becomes even more marked, as we'll see later. For the moment, let us investigate the chemicals of which we are all made.

The chemicals of life

Suppose we could dismantle a human body into its atoms, and sort them out element by element – quite a task, as each of us is made of over a thousand million million million million atoms. For a start, we would find that more than half the atoms are hydrogen, the simplest element. Our bodies contain so much hydrogen that if it was in its usual form of a gas, a balloon filled with this amount would be able to lift us off the ground.

Next in abundance is oxygen. Because oxygen atoms are sixteen times heavier than hydrogen, oxygen makes up most of the body's weight – about three-quarters in all. Third comes carbon, constituting one-tenth of the body's weight; that's enough, if it were in the pure form of graphite, to fill three thousand 'lead' pencils. Nitrogen, the most abundant gas in air, comes next, followed by small amounts of phosphorus, iron, calcium, sulphur and other elements, all making up a tiny proportion of the body's weight, yet essential to its smooth running.

The four most important elements to life – hydrogen, oxygen, carbon and nitrogen – are also the most abundant in the Universe as a whole (if we exclude the unreactive gas helium): we are truly children of the cosmos. All living creatures on Earth have a very similar balance of elements, and this is just a reflection of the fact that all life we know is built up from very similar cells.

So let us go one stage up from atoms, and look at the molecules they form. After all, the sum total of the handful of elements listed above is not a human body, any more than a pile of bricks, mortar, wood and tiles is a house. It is the interactions – the chemical bonding and the reactions – between atoms which gives life its unique properties. No longer do scientists believe that a 'vital force' is necessary to make an organism live: the intricate structures of the molecules involved in life, and the fantastically complicated chemical reactions in the cell, mark out life as an activity very different from anything found in the inanimate world; but the atoms and forces involved are the same.

Indeed the only force directly involved with life processes is the electromagnetic force. The chemistry of life is not concerned with the tiny scale of atomic nuclei where the two nuclear forces operate, while the force of gravity is far too weak to affect the chemical reactions inside the cells. What holds the atoms together in a molecule, and guides the course of the reactions which change one molecule into another, are interactions between electrons, moving in the electric field of their nuclei – all manifestations of the electromagnetic force.

Isolated, uncharged atoms have enough electrons to be

A crystal of Vitamin A, a chemical essential to the normal functioning of human cells. The chemicals involved in the processes are basically no different from any others, only more complicated.

electrically neutral, the negatively charged electrons in orbit equalling in number and cancelling out the positively charged protons in the nucleus. But in most elements these electrons do not fill all the orbits which are allowed by the quantum theory. To fill these extra orbits without losing their electrical neutrality, the atoms adopt the cunning stratagem of sharing electrons with their neighbours. Adjacent atoms are thus linked firmly together by a shared pair of electrons – one donated by each atom – and form part of a molecule.

In the case of the simple gas methane (marsh gas) we have four hydrogen atoms linked to one central atom of carbon thus:

$$\begin{array}{c} H \\ | \\ H-C-H \\ | \\ H \end{array}$$

Carbon has four electrons in its outermost set of orbits, while the quantum laws allow it another four to complete the set. Sharing one electron from each hydrogen atom, the carbon atom finds itself surrounded by its total allowed number of eight electrons – and yet, as each electron spends only half its time with the carbon atom, the average negative charge around the carbon nucleus is still only equivalent to four electrons. A hydrogen atom's complete set is two electrons, so each hydrogen atom in methane is satisfied with its own electron and one of the carbon's four. Four hydrogen atoms surrounding one carbon atom thus form a very stable arrangement, a molecule.

Each pair of electrons shared by two neighbouring atoms constitutes one chemical bond. So carbon can form four bonds and hydrogen one. The electronic structures of oxygen and nitrogen atoms dictate that they form two and three (or, under certain circumstances, five) respectively. It's easy to see how these permitted numbers of chemical bonds can give us simple molecules: one oxygen and two hydrogen atoms make water (H—O—H); one nitrogen and

Diamond is one form of pure carbon, in which the atoms are tightly bonded in a strong lattice, making it the hardest substance   *found in nature. This rough diamond, slightly coloured by impurities, is exactly as it was recovered from the mine.*   *The lattice of another form of carbon, graphite. The layers of atoms are only loosely bound,*   *like a stack of paper, so graphite is soft and finds uses in pencil 'leads' and as a lubricant.*

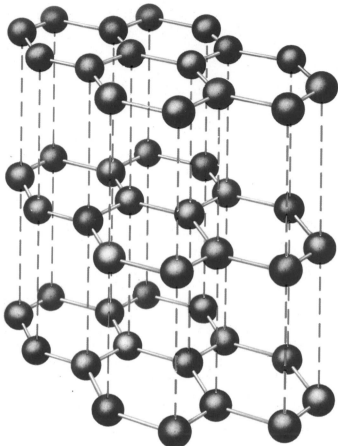

three hydrogen, ammonia

$$(H-N\underset{\displaystyle H}{\overset{\displaystyle H}{}});$$

one carbon and two oxygen atoms, carbon dioxide (O=C=O) – in the last compound, the carbon atom shares two electron pairs with each oxygen atom to form double bonds.

All the molecules in the living cell are made up in this way, from atoms connected by chemical bonds. For simplicity the formulae of the molecules are usually shortened: water is H_2O; ammonia NH_3; and carbon dioxide CO_2. The order of the atomic symbols is traditional for these well known compounds: water could just as well be written OH_2, and ammonia H_3N. For larger compounds, however, it's only sensible to give the symbols in a way which corresponds to their structure. Ordinary alcohol has the structure

$$H-\overset{\displaystyle H}{\underset{\displaystyle H}{C}}-\overset{\displaystyle H}{\underset{\displaystyle H}{C}}-O-H$$

(each '—' representing a pair of shared electrons), and it could be written C_2H_6O; but this formula also applies to

$$H-\overset{\displaystyle H}{\underset{\displaystyle H}{C}}-O-\overset{\displaystyle H}{\underset{\displaystyle H}{C}}-H,$$

a member of the ether family, which would be distinctly unwelcome in a party punch! The formula of alcohol is thus written CH_3CH_2OH, or C_2H_5OH, and this ether molecule as CH_3OCH_3.

Carbon: life's essential atom

This brief digression into the way chemists describe and abbreviate carbon and other compounds as formulae is a background to the way in which the complicated molecules involved in life are constructed, and how biochemists can record their findings in fairly simple terms. And it's no coincidence that carbon compounds have been emphasized so much, for carbon is the element of life. Among all the hundred or so known elements, carbon is chemically unique. Only carbon atoms combine with themselves to form long chains – extended backbones of atoms to which other atoms can attach themselves, to form a huge variety of compounds. The largest molecules in living cells contain several thousand million atoms – and the most important fact of all is that every single one of these atoms is relevant to the continued good health of the cell. After all, a grain of sand consists of many more atoms than this, all carefully arranged in a neat latticework to make a crystal. But despite its size, the giant molecule of a sand grain is very simple, being just a basic unit of one silicon and two oxygen atoms repeated over and over again. By contrast, the big molecules in a living cell are not just repeated, identical arrangements of atoms all the way along: every part of a molecule is different from every other part, and these differences are essential to the complexity of a living organism.

Because these complicated carbon compounds are found inside living cells, but don't occur naturally outside, the chemistry of carbon compounds was originally known as *organic* chemistry, a name which it bears to this day. The vast multitude of reactions which are possible with the wide range of carbon compounds has meant that the chemistry of this single element is studied at most universities in an organic chemistry department as large as the inorganic chemistry department devoted to the chemistry of all the other elements.

The diversity of carbon compounds is essential for life as we know it. Think of all the information you would need to describe yourself as distinct from any other person, or from another living thing: when a human is conceived this vast amount of detail – from the number of toes on the feet to the inner workings of the liver – is carried in code form in a single cell, by just a few enormously long molecules. And despite their size, these molecules are quite stable: the hereditary information can be passed on to one's own children decades later with very little change (though, of course, it is intermingled with information from the other parent). Molecules which are this resistant to change are important to life in other ways, too. The organic molecules which form the outer layers of cells must form a strong bag to keep the cell intact, and resist attack by chemicals from outside. The specialized cells which form our skin, or those that make up the bark of trees, are adapted to producing such tough molecules in quantity.

Yet life needs molecules which can change, in the right circumstances. The energy involved when a seedling pushes through the ground, or an athlete throws a discus, comes from molecules altering – a chemical reaction, in other words. A human takes in a 'fuel' (for example, sugar) and this is broken down in the muscle cells by reaction with oxygen, to produce carbon dioxide and water and, in the process, to release energy to power the muscle. This is not a straightforward reaction: dozens of different molecules are involved in a wide range of reactions. Here carbon compounds must be adaptable; if they were completely inert chemically, life's processes would be impossible.

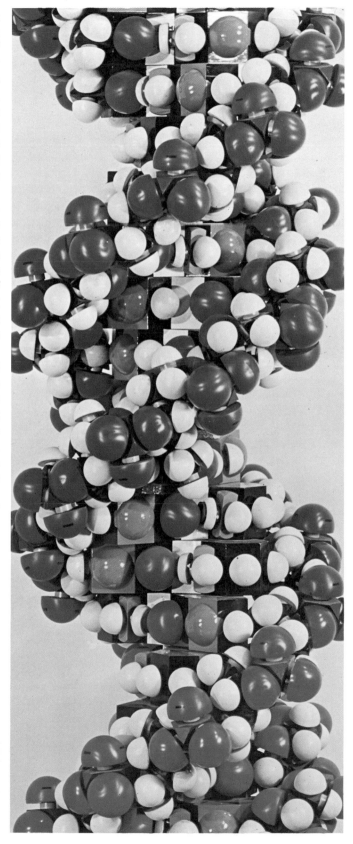

The DNA molecule (left) is replicating by splitting into its two strands. Each strand attracts complementary small molecules from the surrounding fluid, and enzymes weld these together to create two complete DNA molecules, each identical to the original.

So carbon is the ideal element for life. It is common in the Universe, and its compounds are astonishingly diverse, with just the right balance of stability and reactiveness. If we find life elsewhere in the Universe, the odds are that it will be made up from molecules based on strings of carbon atoms.

The cell's control unit

Let's begin our investigation of the cell right at the centre, where a small, round body called the nucleus (not to be confused with the nucleus of an atom) acts as the cell's control unit. Here are stored the 'blueprints' detailing how the cell is constructed, and how chemical reactions are to occur throughout the cell to keep it functioning. The nucleus co-ordinates the cell's activities to ensure its survival, and gives the signal when it is time for the cell to reproduce itself – which cells do in the simplest possible way: by splitting in half into two identical, but smaller, cells.

The cell's nucleus is also a vast 'library': within the nucleus of each cell in the body are the plans for the entire body. These complete plans were necessary to the original single cell from which the body began to grow at conception; and they have been faithfully handed down to all subsequent cells. So each cell nucleus carries far more information than it needs for everyday functioning; and it is theoretically possible to take any body cell, and to coerce it into growing like the single fertilized egg cell present at conception. The result would be another human exactly like the donor of the cell. A controversial American book has recently claimed that a wealthy eccentric paid a doctor to do exactly this, and that he now has a son who is a 'carbon copy' of himself. Although many scientists have denounced this particular claim, they agree that such a possibility (the process is known as *cloning*) is not impossible with our present knowledge of the cell's workings.

The 'blueprints' in the cell's nucleus are in the form of a most unusual molecule, *deoxyribonucleic acid* – a cumbersome mouthful of a name which biochemists prefer to shorten to DNA. The immensely long molecules are coiled around in such a way that they fit neatly into the cell's nucleus. If we could tease out all the DNA in a single microscopic human cell, and put it end to end, the total length would be about a metre (or more than three feet).

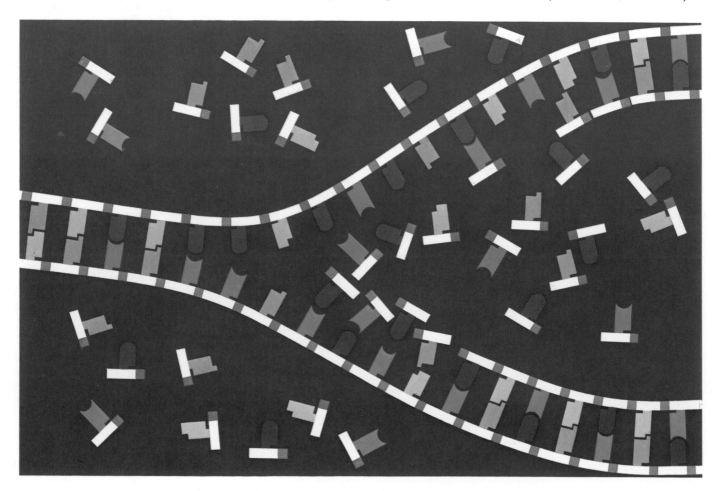

Top:
An electron microscope photograph of a gene, the basic unit of inheritance. It consists of a long DNA molecule, coiled around itself to reduce its length, and so increasing its width.

Below:
A schematic view of DNA, showing the pairings of the 'teeth' carrying the genetic message, and the double helix of the 'backbone'.

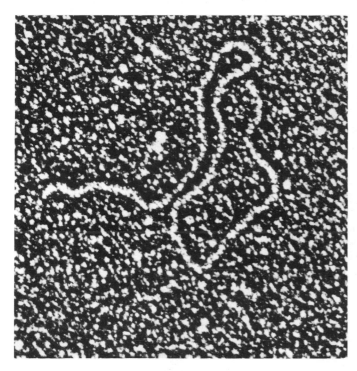

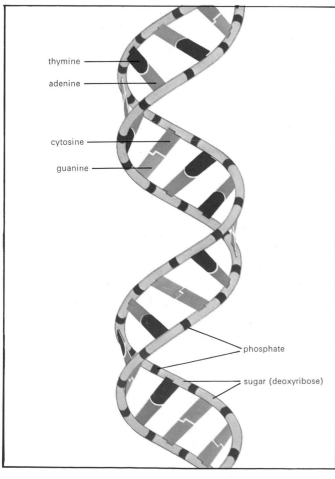

thymine
adenine
cytosine
guanine
phosphate
sugar (deoxyribose)

But how is information about ourselves recorded in the DNA molecule, how is it decoded, and how can it be passed on so precisely to 'daughter' cells? The first big breakthrough in understanding this mechanism of life came in 1953, when James Watson and Francis Crick discovered that the atoms in the DNA molecule are arranged in a 'double helix' – like a lamp cord with each wire twisted about the other. Theirs was an imaginative leap forward, for it is impossible to see this sort of detail in a molecule directly: DNA is far too narrow a molecule to see with a conventional microscope, and even the more powerful electron microscope shows it simply as a dark ribbon. Crick and Watson needed radiation with a wavelength short enough to get right in among the atoms of the molecule: X-rays were the answer. But X-ray scattering techniques can't give clear photographs of the molecule – just a sort of jumbled-up picture, in rather the same way that a hologram is an apparently meaningless pattern of dots until you 'interpret' it by shining a laser beam through it. Watson and Crick narrowly beat their colleagues in deciphering the X-ray results on DNA, and they reaped the laurels of fame – including a Nobel prize – for their feat.

A good analogy for DNA is an immensely long zipper, twisted all the way along. The strips of cloth at either side, which form the two backbones holding the teeth in place, are composed of a repeating group of atoms. These groups are really identical small molecules, which have joined together in their millions to make up the long backbones: each group is a combination of phosphorus and oxygen atoms (phosphate) with an organic molecule called deoxyribose. This last substance is termed a sugar, for it is chemically very like 'ordinary' sugar – sucrose to the chemist – and other types of sugar such as glucose. So these phosphate-sugar groups repeat over and over again, making a stable backbone on either side of the zipper's teeth.

The information content is carried by the teeth. There are four different kinds of tooth, each a group of ten or so **atoms, and like many** biological molecules and groups they have tongue-twisting names: *adenine, cytosine, guanine* and *thymine.* Biochemists shorten these to A, C, G and T. (But note that these shorthand symbols are not the same as the chemists' atomic symbols: C here means the group of twelve atoms making up the cytosine group. Generally it's obvious from the context which shorthand is meant.)

Although these four types of group link the two backbones, the zipper analogy isn't exact. The teeth don't fit between each other, but end to end. They have strong preferences, too. An adenine tooth on one backbone will link only with a thymine tooth on the other, and cytosine only with guanine. So there are two types of link: A–T and C–G, and because of this complementarity, a sequence of, say, AACTG on one backbone must be matched by the

77

The genetic message of DNA is carried out of the cell's nucleus by messenger RNA. At the 'decoding centre', a ribosome, protein molecules are assembled by transfer RNA molecules which match the appropriate sequences on the messenger RNA – each type of transfer RNA carries a different amino-acid.

sequence TTGAC on the other side.

And so one of the mysteries of DNA can be resolved, namely how exact copies are passed to daughter cells. When a cell is about to split, each DNA molecule in the nucleus begins to unzip. As each tooth is exposed, it picks up a partner from amongst the 'loose' teeth floating in the nucleus: an exposed T will pick up a loose A, for example. and the loose teeth each have a segment of backbone already attached; as they line up along the unzipping DNA, busy little molecules called *enzymes* weld all the segments of backbone together to make a continuous new half to complement each existing half-zipper. The result is two new zippers for each old one. And there is no room for error. Each tooth is so set on a particular type of partner that the new half-zipper must be an exact complement of the old half that it is forming against. So this unique molecule is ideally suited to carrying the instructions of life.

But how are the instructions carried? The answer seems too simple to believe: the code of life is simply the order in which the different kinds of teeth occur on each backbone. Every group of three successive teeth holds a particular message: the sequence AAT tells the cell something dif-

ferent from ATA, for example. The actual decoding takes place in special bodies in the part of the cell lying outside the nucleus. And the cell has evolved a clever way of conveying its instructions from the DNA, which stays safely inside the nucleus, to these decoding centres: it sends an acid messenger.

The messenger is a molecule very like DNA, except that its backbone contains a slightly different sugar, ribose, in place of deoxyribose. It's thus called *ribonucleic acid*, or RNA. (There is also a minor difference in the chemicals forming the 'teeth', which can conveniently be ignored here.) To send a message, a small segment of the original 'blueprint' DNA unzips, and attracts complementary teeth, each bearing part of a *ribose* backbone. These are welded into a length of 'messenger' RNA whose teeth match just a tiny part of the long DNA molecule. When the message is complete, the messenger RNA splits off the DNA, and heads into the outer part of the cell, while the DNA zips up that small segment again, safeguarding its vital information.

Decoding the message of life
Scattered throughout the outer part of the cell are thousands

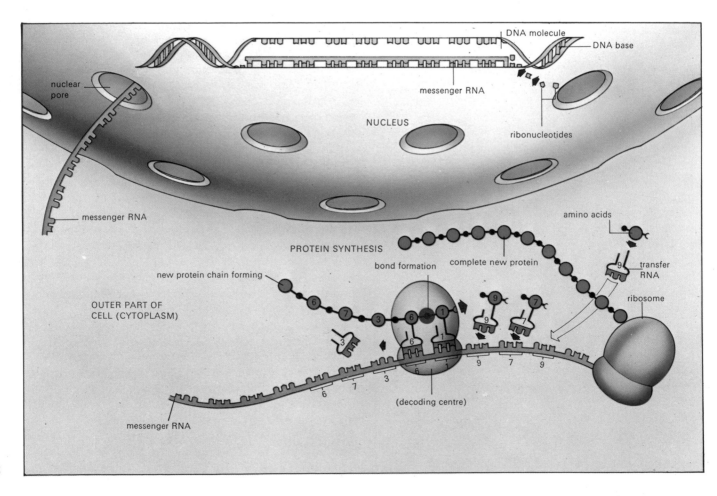

of decoding centres (ribosomes). But before we can understand the decoding process, we must meet the second major kind of molecule (after the nucleic acids DNA and RNA) which is essential for life. This group is the *proteins*, which are also long molecules composed of many smaller units; but there the similarity with DNA ends. The units of a protein are simply strung together end to end: it has the simplicity of a string of sausages, as opposed to the complicated backbone-and-teeth structure of DNA.

The individual 'sausages' are also quite different from the units making up nucleic acids. They are the so-called *amino acids*, organic groups mainly of carbon and hydrogen atoms, but with oxygen and nitrogen atoms near one end. These latter are vital, for they ensure that the chemical bonds between the amino acids joined together as proteins are tough, but not too strong to resist all change. (Incidentally, the word 'acid' in amino acid and nucleic acid doesn't mean that these are particularly corrosive substances; it just denotes a certain chemical similarity to strong acids.)

Living cells contain just twenty kinds of amino acid, but when these are strung together in different orders, in chains of different lengths, the total number of possible proteins is immense. Virtually all of life's processes are carried out by proteins; they are the vast army of willing workers who follow instructions from the DNA molecules in the nucleus. If we look at the constitution of the human body, not in terms of atoms but of the kinds of molecule they are found in, after the surprising first discovery that most of the body's molecules are water (each of us is three-quarters water by weight; and since most of a water molecule's weight is contributed by its oxygen atom, it is the abundance of water which leads to the large weight of oxygen in the body) the vast majority of the remaining molecules are found to be proteins.

Proteins (together with molecules of fats) make up the tough outer coating of cells, including the horny matter which accumulates as hair and fingernails. Smaller protein chains float around in the watery interior of cells, and it is their reactions which are the reactions of life. As well as taking part in reactions themselves, proteins are often more important in controlling and promoting reactions between other molecules. It is in this role of mediator between two otherwise reluctant molecules (the equivalent of the process of *catalysis* in inorganic chemistry) that proteins are known as *enzymes*. It is an enzyme which welds together the backbone of forming DNA molecules; and virtually every chemical reaction in the cell is helped by an enzyme especially tailored for the job.

The messages from central DNA relate specifically to manufacturing proteins from the raw amino-acid material in the cell. They say: assemble so many amino acids in such and such an order; and the result is one specific protein for doing a particular job in the cell. The basis of the DNA code is successive groups of three 'teeth', as described earlier. The message is transported out as complementary teeth by messenger RNA, and when it gets to the decoding centre we find that code breaking is elementary: each set of three teeth codes for a particular amino acid. So the sequence CCT (in the original DNA; it becomes GGA in the messenger RNA) means 'the next amino acid to add is glycine'; GTA means 'add histidine'. The message begins with a 'start' sequence – TAC or CAC – and ends with 'stop' – ATT or ATC. (There are actually 64 possible ways of arranging four different kinds of teeth in trios, and only twenty amino acids, so the cell has inbuilt 'synonyms': both GTA and GTG code for histidine, for example.)

The workers at the decoding centre are short, L-shaped lengths of a third kind of nucleic acid, called 'transfer' RNA. The end of the long branch of this molecule's 'L' has three teeth to match the teeth on the messenger RNA, and the other end has a chemical structure which links onto the amino acid appropriate to its set of exposed teeth. The messenger RNA slowly slides its long molecule past the small decoding centre itself; and as each set of three teeth moves in, it is recognized by the appropriate transfer RNA, which moves in and locks teeth temporarily. The amino acid it bears at the other end is then welded on to the growing protein chain by specialized enzymes. As the messenger RNA moves on, the L-shaped transfer RNA, now stripped of its amino-acid, is rejected, and another arrives which matches the next three teeth and delivers the next amino acid. Thus the nucleus' message is delivered safely, and the required protein is manufactured.

With the breaking of the cell's code, scientists believe that the basis of life's processes is now understood. But the interlocking reactions are so complex that the complete picture is still a distant dream. Recent research has shown, for example, that considerable portions of the DNA's message are often left out when messenger RNA is assembled. How does it 'know' which parts to leave out? There must be side reactions, even in this straightforward copying process. Yet scientists now know where and how life's information is stored; how it is faithfully passed on to daughter cells; and how the code is translated into proteins by the cell.

Some of the implications are almost miraculous – or disastrous, depending on the point of view. It is possible to cut up the DNA in cells and rejoin it to change the messages it carries, and to insert bits of DNA from one cell into that of another. These techniques of 'genetic engineering' (a *gene* is the total inherited information carried by a length of DNA) have the capacity for both good and evil. Insert a length of DNA which orders the production of a

useful protein – perhaps insulin, which is essential for diabetics – into a simple cell which reproduces rapidly, and in no time at all you could have a vat full of cells churning out insulin in quantity. And at a deeper level, careful experiments in changing around DNA can tell us far more about inheritance, and help us to understand inherited diseases.

But there are dangers, too. Most bacteria are harmless, and many live quite happily in our digestive tracts – including the genetic engineers' favourite, *Eschicheria coli*. Suppose an experimenter changed the DNA of just one *E. coli* cell so that it became lethally toxic; and that this cell escaped from confinement. It could then multiply indefinitely, donating its fatal trait to its offspring. In this admittedly extreme scenario, an epidemic of a new disease, with no known cure, could be the horrific result of a moment's carelessness in the laboratory. In view of dangers like this, new safety measures have been introduced at genetic laboratories all around the world in the past few years, and scientists involved in this kind of research have been under pressure from concerned members of the public.

Fuelling the cell
A living cell cannot function without an energy supply. And in looking at the way a cell gets its energy, we meet the great divide which separates animal and plant life. All the other cell processes – the reproduction of DNA, the synthesis of proteins and its associated code – are identical in all forms of life, from single-celled bacteria to human beings; from elephants to the common rhubarb plant. But plants and animals get their energy in different ways.

Animal cells are powered by burning sugar. The energy we use comes mainly from chemicals like ordinary sugar, either included in our diet, or from broken-down fat molecules in the cell. And burning is exactly the right word. Hold a sugar cube in a gas flame, and after melting and charring it will burn away entirely as carbon dioxide gas and water vapour. An animal cell absorbs sugar and oxygen from outside, and also converts them into carbon dioxide and water – with a release of energy in a rather less spectacular way.

The difference between burning sugar in a gas flame and in our cells is obviously a matter of temperature. Either way, burning sugar gives out energy. A bowl of sugar at body heat left standing in air doesn't spontaneously change into carbon dioxide and water, because a certain amount of energy must be put in to start the reaction. A flame will do, but it is unnecessarily drastic. In our cells we have enzymes once again to thank for 'burning' sugar at our relatively low body temperature. They control and smooth the way for a whole series of reactions by which sugar is successively broken down into simpler molecules before it ends up as carbon dioxide and water.

These reactions take place within hollow, tube-like 'power houses' in the outer part of the cell. The enclosing membrane of this little body becomes charged up in the final stages of sugar burning – the total electric current charging up all the miniature 'power stations' in our bodies is enough to light continuously a 100-watt electric bulb! This energy is stored as the difference in voltage between the inside and outside of the power houses, like the difference between the ends of a battery. But the cell can't use electric energy directly. It needs a chemical 'currency' to transport energy to wherever in the cell it is needed.

The molecule it uses is ATP (*adenosine triphosphate*). This is essentially one of the loose teeth which can join into a chain of RNA, the tooth itself being adenine (A); and it carries a ribose-phosphate backbone group. In ATP there are two extra phosphate groups, making three in all. The final phosphate group is only loosely attached: 'trigger' the ATP molecule and it will release this phosphate group with a burst of energy. The molecule that is left migrates back to the power houses, where the electric energy forces another phosphate group on. The reconstituted ATP can then wander off through the cell until its energy is required.

Plant cells, too, use ATP as their energy currency, but they make it in the opposite way. They take in carbon dioxide and water, and from it build up sugars while excreting oxygen. Plant cells are facing an uphill process here; they must take in extra energy from outside, and they get this from sunlight by the almost miraculous process of *photosynthesis*. The plant cell's small power houses evidently need to be different from an animal cell's; and one obvious distinction is in colour – it is the *green* powerhouses within the cells which give plants their characteristic colour. A molecule called chlorophyll absorbs photons of red light to start the photosynthesis process, and the reflected light, being white light minus red, looks green.

Each light photon ejects an electron from a chlorophyll molecule; and this escaped electron has enough energy to start off a whole series of reactions. The upshot is eventually to charge up the outer membrane of the power house, and this electrical energy is once again used to put together ATP molecules. In both plant and animal cells, an energy source charges up a membrane, and from here on the process of energy production and transport is the same.

So even though animals take in plants' 'waste products' of oxygen and sugars, while plant cells thrive on the animals' wastes of carbon dioxide and water, there is surprisingly little difference in their cells' internal mechanisms – in the DNA 'blueprints', in the decoding process, in the types of amino acid used in proteins, or in the later stages of energy production.

There is an astonishing unity among life-forms on Earth.

Life's place in the Universe

The complexity of life's processes is staggering. In the last few sections we've seen just a few of the complicated molecules and interlinked chemical reactions which give life its unique properties. Yet it is not so much the complexity of a living cell as its organization which is important. A lump of granite appears as a complex mixture of tiny crystals when seen under a microscope, but they are not organized – we could rearrange the crystals and the rock would still be granite. Alter the internal structure of a living cell, however, and it would no longer be able to carry out its life-giving reactions. In multi-celled creatures like Man, the organization is carried to further extremes, for it is the interconnection of ten thousand million nerve cells in our brains which allows us the capacity for thought.

And so, although we can explain life's processes in terms of fundamental particles and the electromagnetic force, and also (in the next section) show how it can arise in the astronomical evolution of the Universe, life is really a separate facet of the Universe. As well as investigating the very small and the very large, there is a frontier of knowledge connected with the more abstract notion of the development of organization. It is very easy to be impressed by extremes of size, big or small; but it can be misleading. The Sun is a huge and powerful object, but it is really very simple: an astronomer could write down all the essential facts about the Sun on one page, and with this information we could (theoretically) build a star exactly the same. But the information needed to make one microscopic cell would fill several encyclopedia sets. Without being too immodest, we can claim that the most marvellous objects in the Universe are not quasars, black holes or quarks, but we ourselves.

With that thought, let us go back to our own origins, the beginning of life on Earth.

Life from non-life

The surface of the early Earth was an artist's impression of Hell. Some 4,000 million years ago (600 million years after it first began to form) the Earth had grown to about its present size, but in no other way would we recognize it. Huge rocks fell from the sky as Earth swept up the final fragments of debris near it in space, and their impact blasted out enormous craters hundreds of kilometres across. Radioactivity was heating up the Earth's rocks, causing huge sheets of molten red-hot lava to well out and spread over the surface; while the noxious gases from volcanoes shrouded the Earth in a thick atmosphere, laden with clouds and rent by continuous thunderstorms. Soon afterwards the rains came. The vast amount of water vapour exhaled by volcanoes condensed and fell as a continuous downpour on the Earth's sizzling surface, cooling it down, and eventually

The first stage in the growth of a complex creature: the fertilized cell splits in two. This rat egg-cell has split two days after

fertilization, and further cell division will turn these cells into a living, breathing rat in only 20 days.

The hereditary material of man is packaged into 46 chromosomes in each cell nucleus (the dark blob is the nucleus of another cell). The reproductive cells each

contain only 23, so a fertilized egg regains the normal complement, half from each parent. In this way, variety can occur in the offspring.

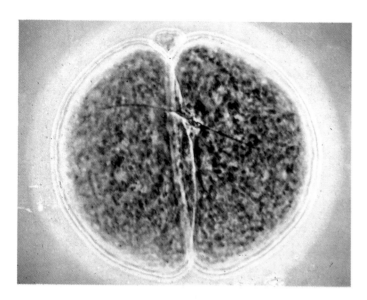

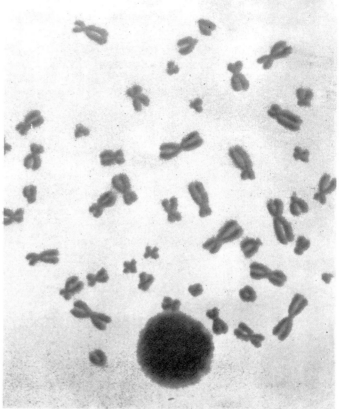

collecting as wide oceans.

Not the kind of place we would choose for a holiday; but these unfriendly conditions were ideal for our earliest ancestors. Life must have begun from non-life, and today's apparently benign conditions are not right for this act of genesis. If the type of complex organic molecules found in our cells were exposed to the oxygen-rich atmosphere of today, they would eventually break up into carbon dioxide, rather than amalgamate to produce living organisms. Any-one who has seen a forest fire knows only too frighteningly well how unstable the chemicals of life are in the presence of oxygen. After all, it is from the slow burning away of organic compounds by breathed-in oxygen that animals get their energy; the gaunt look of a starving man is a result of the literal burning away of his body molecules in a desperate attempt to provide energy.

On the other hand, Earth's first atmosphere was without free oxygen, while it was rich in volcanic gases: carbon dioxide, carbon monoxide, hydrogen sulphide, ammonia, hydrogen cyanide. Poisonous as these gases are to us, they were the breath of life to the first living things. In the fierce environment of the primaeval Earth, these compounds reacted together to make up new molecules, organic chemicals larger and more complex than the original gases.

Such ideas are not just elegant theory. In a famous experiment back in 1953, American chemist Stanley Miller prepared a mixture of gases similar to those of the Earth's early atmosphere, and flashed a 60,000-volt spark – artificial lightning – through it for a week. To everyone's astonishment, the liquid left at the bottom of the flask was a rich soup of organic compounds, including four amino acids. Spurred on by Miller's results, other scientists have investigated different gas mixtures, and subjected them to other disturbances which must have rent Earth's first

atmosphere. These included shock waves ('sonic booms' from falling asteroids and meteorites), ultra-violet radiation (from the Sun) or simply intense heat (of the molten rocks). All these experiments produced the simplest molecules of life, and in profusion. The 'primaeval ocean' at the bottom of the laboratory flask contains all the amino acids necessary for building proteins; the individual units from which DNA and RNA are built up; and many other compounds, some of which are found in today's cells, and others for which the life process evidently had no use.

So the production of small organic molecules seems almost inevitable on the early Earth. Although this was a surprise in the 1950s, modern astronomy shows us that organic molecules are surprisingly common in the Universe. Radio astronomers pick up signals from distant gas clouds at the wavelengths broadcast by various organic molecules. Some fifty different molecules have now been identified in space, and although none is very complex, their mere existence shows that organic chemistry does go on in the gas clouds between the stars – a possibility unsuspected twenty years ago. These reactions probably occur on the surfaces of tiny dust grains in the cloud; and it's an intriguing possibility that more complex molecules than those yet found may exist on the grain surfaces.

Astronomers also find organic molecules in some of the

meteorites which fall to Earth. As we saw in a previous chapter, these 'carbonaceous chondrites' are the earliest material of the solar system, unaltered for 4,600 million years. Such meteorites have only recently been investigated in detail, because although they are very common in space, few of them survive the fiery passage through the Earth's atmosphere. Once they land, these crumbly rocks are quickly weathered away unless they are seen to fall and are immediately taken to a laboratory.

On the last Sunday in September 1969, a bright light flashed across the morning sky near Melbourne, Australia. Over the small town of Murchison, the meteorite exploded with a thunder-like boom, and fragments of rock rained down. A total of 82 kilogrammes (more than 180 lbs) was collected and taken for analysis – and among the few per cent of carbon compounds, scientists found several kinds of amino acid, the components of protein. Somehow, the basic units of life had assembled out in space, either during the coming together of the solar system, or even before that, in the gas cloud from which the planets eventually condensed.

There's no difficulty in understanding how organic molecules up to the complexity of amino acids, or the groups which make up the nucleic acids, could have occurred on Earth. Some may have survived from the fall of carbonaceous meteorites, while huge quantities must have been synthesized from the early volcanic atmosphere. The early oceans were a 'primaeval soup', with the consistency of thin chicken soup, according to biochemist Leslie Orgel – although it would have smelt like a chemistry laboratory.

The next stage in life's story is the most obscure. Over the course of 500 million years, these simple molecules must have come together to make up DNA, RNA and proteins – but this is only a part of the story. Large molecules can readily be assembled from smaller ones in the laboratory, but the result is not life. The interdependent chemical reactions had to evolve too. Enzymes (proteins) 'unzip' the DNA to allow it to reproduce, and to expose its message for decoding; and conversely this message is a plan for the production of proteins. Somehow a series of reactions between simple nucleic acids and simple proteins must have gradually increased in complexity, and in the process produced larger and more complex molecules: life processes lifted themselves up by their own bootstraps. And the energy-producing mechanism, relying on enzymes as well, must have grown up at the same time. One thing was readily available for all these reactions: as a source of raw materials and 'fuel' they were living in the bountiful medium of the life-giving primaeval soup.

The earliest fossil cells are found in the Fig Tree cherts of Swaziland (but these are just the earliest to have left traces, which most single-celled organisms would not do). Dating back 3,400 million years, these simple organisms show that life began very early in Earth's history. Let's scale down the Earth's past, to a mere 46 years instead of 4,600 million. We can say that it spent its early childhood under continuous battering from rocks falling in from space; but by the age of seven, Earth's crust had formed, and simple organic molecules were beginning to form in the atmosphere. By its eighth birthday, at the latest, the rains had come and Earth was awash with oceans – the oldest rocks we know date from this time, and show signs of water action. And in the next four years, the chemicals of life had formed themselves into the first self-contained cells, each an enormously complicated chemical system. So Earth has borne life forms from early adolescence right up to its present middle age, some 34 years out of a lifetime of 46.

Very early on, these primitive cells ran into a crisis. After they had gobbled up the organic ingredients of the soup around them, there was no obvious source of food. Only carbon dioxide in the atmosphere and dissolved in the sea held out any hope. The development of photosynthesis rescued the living cells from eating themselves to extinction, for once they could use sunlight to turn carbon dioxide and water into organic molecules, they were assured of a long-lasting energy supply.

But the by-product of photosynthesis is oxygen. Over thousands of millions of years, this 'waste' from plant cells gradually replaced carbon dioxide in the atmosphere. And so Earth's atmosphere slowly altered. The initial thick carbon-dioxide blanket had been thinned out as this gas dissolved in the seas and reacted with exposed rocks, and now the remainder was being broken up by plant cells. The overall result of photosynthesis was to put the carbon atoms from the gas into living creatures, and leave the oxygen in the atmosphere. Oxygen, being a reactive gas, soon broke up the other noxious gases remaining from the primaeval volcanic atmosphere. Relatively inert nitrogen was an exception: although it started off in only small quantities compared to the overwhelming amount of carbon dioxide, it did not dissolve to the same extent in the oceans; nor did it react with exposed rocks; nor with the new atmospheric ingredient, oxygen. As other gases became depleted and our atmosphere thinned, nitrogen stayed the course to end up as the most abundant gas in air today, making up almost four in every five molecules. One in five is oxygen; while argon, resulting from the break-up of radioactive potassium, lies in third place, constituting 1% of air. Carbon dioxide now makes up only one molecule in three thousand in Earth's atmosphere – contrasting strongly with Venus and Mars, whose primitive atmospheres are still composed almost entirely of this gas.

As the carbon dioxide disappeared from the air, plant cells could have faced a second crisis, choking in their own waste oxygen. Animal cells restored the balance, however, for they consumed plant cells, 'burning' them in oxygen to restore carbon dioxide to the atmosphere. Today's life still maintains this balance, as plants absorb our exhaled carbon dioxide, and we eat their carbon compounds and breathe in their oxygen waste.

Oxygen in the atmosphere produced another great revolution in life, for the rare three-atom molecule of oxygen, ozone, formed a layer at the top of the atmosphere which filtered out harmful short-wavelength ultra-violet from the Sun. No longer needing the protection of sea water as a screen, life emerged on to land. And by this time, plants and animals were no longer single cells.

Change for the better

Life has obviously taken a tremendous step forward since the days when all living things were microscopic cells swimming in the oceans. Plants and animal fossils of increasing complexity begin to appear in rocks a few hundred million years old, and once cells had developed the knack of working together as a multi-celled organism, a huge assortment of different kinds of life appeared on Earth: animals ranging from insects to dinosaurs; and plants such as the enormous 'ferns' of the Carboniferous era, whose fossil remains are now dug up as coal.

But how could life have taken so many diverse paths, when the ingenious 'unzipping' of DNA seems to guarantee that all daughter cells will be exactly like their parents? The answer lies, in the first place, in *mutation*. Radiation from space, or from radioactive elements in the ground, will occasionally disrupt and change the DNA in a cell, subtly altering its message. Experiments with the fast-breeding fruit fly, *Drosophila melanogaster*, and other species show that a mutation occurs only about once in every million DNA copyings – not a bad success rate, if we think of it in terms of gramophone record pressings, for example. Alternatively, since each cell's DNA carries as much information as several encyclopaedia sets, it's equivalent to copying a shelf of books a million times, with only a single mistake.

The vast majority of mutations are harmful: the altered DNA cannot control the cell properly, and it will soon die. But the odd one will just happen to give the cell some advantage over its ancestors and relations; it will thrive at their expense, and its descendants will ultimately survive where the original type of cell could not. The development of photosynthesis in cells must have been the result of several advantageous mutations, ultimately producing cells which could survive when all the primaeval soup was consumed.

Living organisms on Earth are always competing for food and living space: this is the second driving force behind the proliferation of life forms. In the struggle for survival, a creature inheriting a useful characteristic is more likely to reach the age of reproduction, and its descendants – bearing this same characteristic – will gradually replace the original type. So the types – or species – of life on our planet are continually changing. Our present-day species of dogs, elephants or primroses are no more permanent than were the dinosaurs and giant Carboniferous ferns.

Charles Darwin was the first to publish this idea of evolution by natural selection of the fittest, in his famous and controversial book of 1859, *Origin of Species*. Although it may seem obvious enough to us today, his heretical idea that species of plants and animals were not fixed by God at the Creation led him into a head-on clash with the views of the Church – which were also held by orthodox scientists of the day. Darwin himself had no wish to start a controversy, but the bitter religious and scientific dispute went on for decades. Honours were showered on him from all over the world, but the only honour he received from Queen Victoria's government was burial in Westminster Abbey.

By the process of evolution, life emerged from single cells to the complicated life forms of today. A few hundred million years ago – very recently in Earth's history – some cells developed the quirk of living together in permanent colonies. It was an advantage for such cells to take on specialized tasks: in animals, some would digest food, others carry chemicals around the colony – or 'body', as we would now call it – and some would take on the task of locomotion, while other cells became adept at transporting electrical signals. A special group of cells developed to handle the task of reproduction.

For at this late stage in evolution, two new and very fundamental factors appeared on life's scene. Multi-celled organisms invented ageing and sex. The earlier single-celled creatures were effectively immortal: they kept dividing into pairs of daughter cells, each identical (apart from the rare mutation); and, unless they were eaten or ran out of food, the original cell 'survived' indefinitely in the form of its own offspring. Single cells do not age, and they reproduce in the simplest possible way: by splitting down the middle.

But multi-celled organisms have a great problem with reproduction. We can't just let every body cell split in two: there would then be two intermingled creatures. So in both animals and plants, part of the organism is set aside for reproduction. From Nature's point of view, this is the most important group of cells in the body: from here the genetic information we inherit at birth will be passed down to subsequent generations. The rest of the body is just an intricate system for safeguarding and nourishing the sexual organs until they have matured enough to create future generations. And from this point of view, ageing is a very important part of evolution. Each new organism starts out as a single cell, and must grow into an adult over a period of time, as the initial cell splits and splits again to make up the new 'colony'. But if all adults of the species were immortal, numbers would soon grow to the point that they would monopolize food supplies, and new generations would not survive infancy. For the evolutionary process, adults are only necessary up to the time that they reproduce; after this, they must die a natural death and leave the stage clear for their descendants. And so evolved the still little-understood process of ageing.

Each offspring grows from a single cell produced by the parent's all-important sex organs. But the diversity of life can be enhanced by mixing DNA from two parents, and so the subtle process of sexual reproduction has evolved: cells from the two parents merge and combine to make a single cell with mixed characteristics. The DNA 'information' from each parent – be it the colour of the eyes, or the stem height of a plant – must vie in the new cell's nucleus. The science of genetics concerns itself with the question of how one set of information – a *gene* – is selected rather than another.

With the advent of multi-cellular organisms, first in the sea and then advancing onto land, the scene was set for an enormous increase in the number of species of life, and in the pace of evolution. All this is very recent in Earth's long life. On our 46-year reduced time scale, the first cells formed when the Earth was about 12; but multi-cellular life forms did not emerge until a few 'years' before the present time. At 42, plants and animals came up on to the land; while only a year ago, dinosaurs ruled the Earth, and the first flowering plants appeared. A couple of weeks ago, hairless apes began to use tools and fire; and in the past few seconds they have acquired the technology to escape from our home planet Earth in tiny metal capsules. And they have begun to ask: is there other life in the Universe?

Havens of life

The kind of life we know needs a home planet like the Earth – a watery world. Life's chemical reactions take place in water: both the original, seemingly miraculous combination of molecules into the first living cells in the primitive oceans, and the present-day life processes in our own cells. Chemically, water is a very versatile solvent; and our life chemistry depends almost as much on the properties of the water molecule as on those of the carbon atom.

Looking around our solar system, the bleak climates of the other planets seem to show our watery Earth as an exceptional place. Tiny Mercury and the Moon don't have sufficiently strong gravity to hold down an atmosphere, and any water would boil off into the near vacuum of space. Venus probably began as Earth's twin, but it was too close to the Sun: too hot for water vapour to fall as rain; and Venus' carbon dioxide atmosphere muffled it further until its surface temperature reached the melting point of lead. Mars, circling the Sun outside Earth's orbit, is too chilly. Water there lies trapped in the polar caps and in the surface rocks as permanent ice. Although surface channels like dry river valleys show that some water once flowed on Mars, it's not certain that it was widespread enough to foster life; and the Viking experiments found virtually no carbon compounds in the Martian soil.

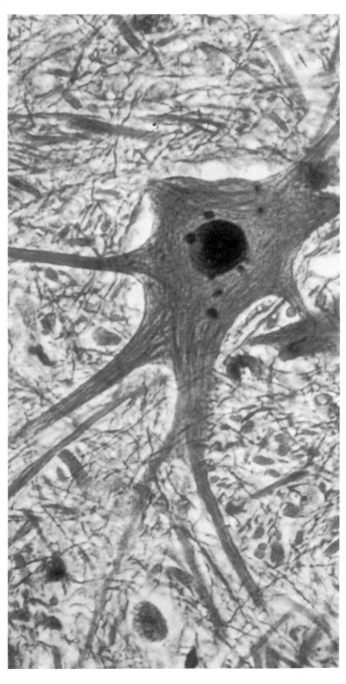

What of the giant planets? In Jupiter's atmosphere, reactions like those on the primitive Earth must make a wide range of organic compounds; but, since it has no oceans or rocky shores, most life-scientists are dubious about whether cells could form. Undaunted, leading optimist Carl Sagan of Cornell University has hypothesized that huge gas-bag creatures may live in Jupiter's atmosphere, floating through the thick gases like fish in water.

But for life akin to our own, we must look outside the solar system. In our Galaxy, there are some hundred thousand million stars – and present astronomical theories show that most will form with attendant planets. Computer scientist Stephen Dole has mimicked planetary formation in a large computer, and he has found that rocky planets like Earth generally form near the star, with gas giants further out. In a reasonable percentage of cases, he ends up with an Earth-sized planet at such a distance from the star that water is liquid – the prime requisite for life processes to begin. And note that even if we are pessimistic enough to think that only one star in a thousand has such a comfortably familiar planet, the vast number of stars in our Galaxy means that we are still dealing with a hundred million 'Earths'.

Such a planet would undoubtedly end up with oceans of organic 'soup', like the early Earth. And here we enter the realm of speculation: does life inevitably follow from such a mixture, or was its occurrence on Earth due to a freak coming together of molecules, which may not have been repeated elsewhere? The remarkable similarity of all life's processes seems to indicate that all living cells may be descended from just one original living cell – perhaps just a lucky coincidence of molecules that would not be repeated on even one other of a million Earth-like planets. But optimists argue that life on Earth began surprisingly quickly, in only a hundred million years or so, and this suggests it followed an inevitable sequence of events. Really, not enough is known of life's beginnings for us to decide – yet.

This argument can be repeated at each of life's crises. Was it inevitable that cells 'invented' photosynthesis when the 'soup' became too thin? That oxygen-breathing animal cells emerged to prevent the plant cells from suffocating? That an intelligent species developed which wants to find out about the Universe? Or were we just lucky?

The optimist could conclude that all these will suitably happen, and that there could be millions of planets supporting intelligent life; the pessimist might end up thinking we are the only intelligent life form in the Universe.

Talking across interstellar space
Among the optimists there are many scientists prepared to take positive action. Accepting that there are other civiliza-

tions 'out there', surely some must be trying to get in touch with other intelligent races. And so, for the last twenty years, astronomers all around the world have been looking out for intelligent signals from space.

Without knowing the first thing about possible extra-terrestrials, our search for signals may seem like groping in the dark: we can only argue from our own experience. Planetary scientists keep in touch with distant probes by means of radio waves, and this was the first possibility that came to mind. In 1960, Frank Drake began the first search for extra-terrestrial signals: his 'Project Ozma' was designed to detect any intelligent communication from the locations of two nearby Sun-like stars. For three months he 'listened in' with the big 26-metre (85-foot) radio telescope at Greenbank, West Virginia; but the results were as insubstantial as the land of Oz, the project's namesake.

Nonetheless, radio is an efficient way to flash messages across the Galaxy. The world's largest radio telescope at Arecibo, Puerto Rico, could receive a message broadcast by a similar telescope anywhere in our Galaxy, and radio waves are not blocked by the all-pervading dust between the stars. But there's the problem of tuning in. As with an ordinary radio set, astronomers trying to tune in to a cosmic signal must know the wavelength put out by the transmitter. And since there's no interstellar programme guide, we must try to decide on a wavelength which would be logical for extra-terrestrials to use.

Frank Drake turned to 21 cm, the natural wavelength of the simplest and most abundant element, hydrogen. Over 600 nearby stars have now been surveyed at this wavelength, with a far greater sensitivity than Drake's original investigation. And the search has been complemented by whole-sky scans, in the hope of picking out more distant super-strong transmissions. All, as yet, has been to no avail.

But is 21 cm the obvious wavelength? Since the Galaxy is filled with hydrogen gas, there's always a background of 21-cm radiation to interfere with observations. Some scientists think that an intelligent race would choose a 'quieter' wavelength: perhaps between 21 cm and the 18 cm wavelength of the hydroxyl ion (OH), which is water with one hydrogen atom missing. Since hydrogen and hydroxyl together make up water, this wavelength band has been dubbed the 'water hole': maybe it is the best wavelength for water-based life to meet at. But then again, water molecules themselves emit at 1·35 cm wavelength, and wouldn't this be a more likely wavelength to choose?

Searches at all these wavelengths are now under way, mainly in America and Russia. No signals have yet been picked up; but there's always the lingering doubt about our tuning. Over the past twenty years, at least three very logical guesses have been made about the 'obvious' wavelength for interstellar radio contact: it's possible that a

civilization a million years more advanced than ours would have a completely different idea.

And they might have reasons for not using radio waves. Infra-red and ultra-violet wavelengths are other possibilities, although they would not travel so far through our dust-laden Galaxy. Herbert Wischnia has already used the large orbiting ultra-violet satellite *Copernicus* to scan nearby Sun-like stars; but he, too, has drawn a blank.

Ordinary light has its drawbacks as an interstellar messenger, for an intelligent signal would be overwhelmed by the brilliance of the planet's parent star, its 'Sun', when seen across the vast reaches of space. But lasers offer a way out, for it would be possible to tune a laser to shine at one of the wavelengths where the parent star shines weakly – in a dark absorption line of its spectrum. If this is the Galaxy's communication band, the first extra-terrestrial signals may not come from a sustained hunt with large radio dishes, but from an astronomer studying star spectra as part of a completely unrelated research project.

Then again, what if everyone in the Universe is just 'listening in'? To be logical, we ourselves should transmit signals for others to detect. In fact, Man has been doing this unwittingly for decades, as domestic radio and television signals have leaked off into space. These signals have already reached the nearest Sun-like stars. Anyone there with a large radio telescope, who happens to hit the right frequency, will already know that technically skilled life exists on Earth (although whether he considers it intelligent may depend on which programmes he picks up). And in November 1974, the huge Arecibo telescope was used to transmit Man's first intentional signal to the stars. The message was beamed to a large, old star cluster in the hope that some of its stars might have planets with advanced life on them. Unfortunately, the message will take 24,000 years to arrive, even at the velocity of light. We should be looking out for a reply in some 48,000 years time. This underlines the incredible loneliness of interstellar space: the vast distances mean that messages must be monologues rather than conversation; and if we do receive a signal, it may come from a civilization which has disappeared by the time we receive it. Or *we* may disappear before we receive it!

Star travel

Man has always wanted to explore. There's far more romance in a manned expedition to the Moon than in a robot mission which could perform almost the same tasks for a fraction of the cost. And so it's likely that, one day, a manned spaceship will head for the stars.

The necessary technology is already with us. The Jupiter fly-by probes Pioneer 10 and 11 are travelling through the realm of the outer planets with a speed sufficient to take them clear of the Sun's gravitation. If they were headed

towards the nearest star, they would arrive in some 80,000 years.

We could build spaceships now which would take men to the stars in this sort of time. The problem, of course, is ageing. Hibernation or freezing are possibilities which have been mooted, although they are not medically feasible yet. A more realistic approach is advocated by Gerard O'Neill of Princeton University, a physicist who pioneered the type of nuclear physics 'storage ring' which was essential to the discovery of the charmed quark. His latest project involves the 'humanization' of space, the establishment of huge, self-supporting space colonies of thousands of people, in orbit around the Earth. And there's no reason why some of these colonies, equipped with a suitable rocket engine, should not head off on ponderous interstellar voyages. O'Neill reckons that within a million years – an incredibly short time, astronomically speaking – the descendants of these first pioneers could spread across the entire Galaxy.

The question of interstellar propulsion is the real problem if we want to get to the stars more quickly. Chemical burning rockets of the kind we see blasting off from Earth today are not efficient enough to propel us to the stars more quickly than the Pioneer explorers. A slow but steady acceleration, maintained for years, is the only answer, and the only really efficient process we know is nuclear energy. A nuclear reactor could heat hydrogen gas in a 'combustion' chamber, and the hot gas could then push the rocket forward, like the hot burnt gases from a chemical rocket. Or the nuclear power could be used to strip electrons from atoms, and shoot the positively charged remains backwards by electrical repulsion, so pushing the ship forward.

The only 'practical' design for a starship so far published literally bombs its way into space. The British Interplanetary Society's (unmanned) probe to Barnard's Star would be propelled by the explosion of tiny hydrogen bombs, ejected behind it at the rate of 250 per second and each triggered in turn by laser beams. The successive blasts would be caught by a magnetic umbrella at the rear of the ship, which would thereby be pushed up to higher and higher speeds. A two-stage craft of this kind could traverse the 5·9 light years to Barnard's Star in 'only' 47 years.

But the real limitation to interstellar flight is Nature's absolute speed limit: the velocity of light. According to Einstein's theory of relativity, nothing can travel faster than electromagnetic radiation; and so the minimum journey time to Barnard's Star is 5·9 years. And this is the nearest star known to have a planetary system: even the optimists reckon that the nearest inhabited planet is unlikely to be nearer than several hundred light years. According to present scientific theory, it will never be possible to just nip off for a quick holiday on another star's Earth, or speed around the Galaxy like a jet streaking over the

Earth's surface.

The speed restriction does seem very fundamental to the Universe. Science-fiction writers can get around it by diving in and out of 'hyperspace', but physicists cannot fit that possibility in with current theories. It's no exaggeration to say that our understanding of the Universe would be turned on its head if faster-than-light travel were found to be possible. Of course, this may happen some time in the distant future, when our present theories would seem as ridiculous as those of the flat-Earthers. Nonetheless, present theories seem to be giving us a more and more unified picture of the workings of Nature, and the limit of light's speed is likely to be with us for some time to come. For a while, at least, our experience of Nature's most complex and fascinating achievement, life itself, will remain tied to the Earth . . . unless somebody comes to visit us?

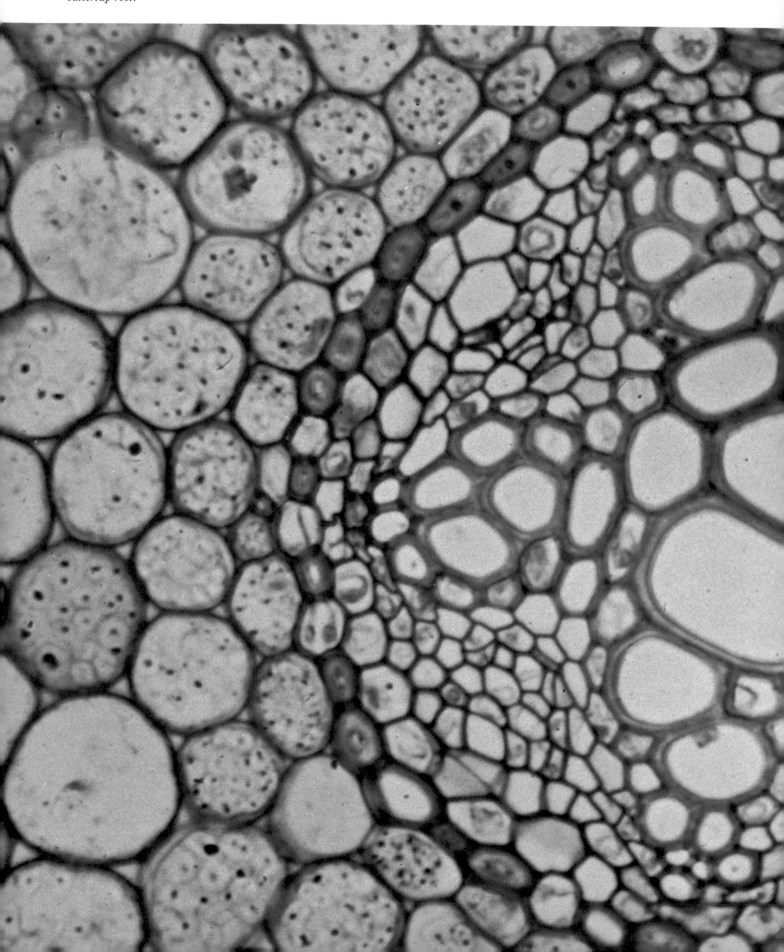

The microscope reveals beautiful symmetries on a scale far smaller than the eye can see. These cells form the central part of a buttercup root.

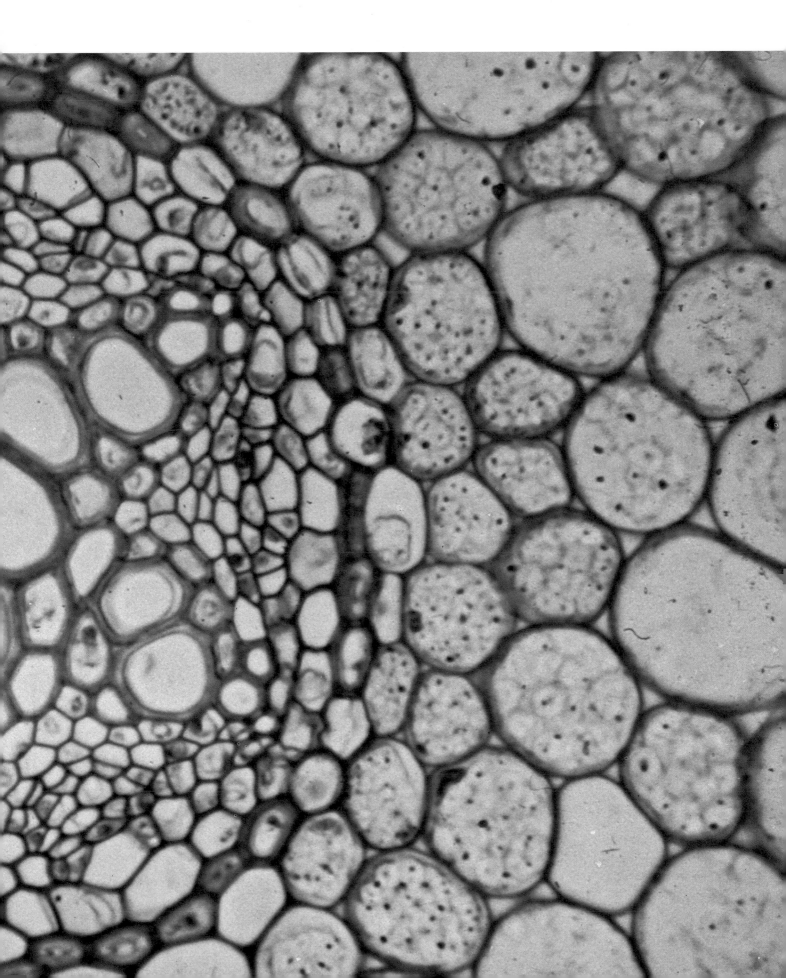

Different methods of illuminating specimens under the microscope emphasize different aspects. This group of diatoms (one-celled algae) are seen in (top to bottom): polarized light;

darkground illumination – in which only obliquely scattered light is seen; and in normally transmitted light, shining through the specimen.

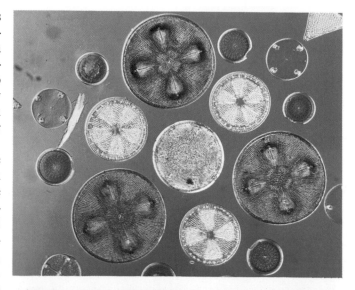

Exploring the very small is as difficult and expensive as exploration of the Universe beyond our planet. Multi-million dollar particle accelerators vie for funds with observatories and space probe missions; top-class, dedicated scientists backed by the latest computers are drawn to both frontiers of knowledge. And, as we shall see, what they find at the limits of the ultra-microscopic defies common sense as much as astronomers' discoveries strain our imagination at the other extreme of size.

Bring this page closer and closer to your eyes, and the detail becomes more apparent. The colour pictures turn out to be merely patterns of dots. But there's a limit, where the straining eye can focus no nearer, and until the seventeenth century, no one knew of anything smaller. Around this time, the simple microscope – effectively just a powerful magnifying glass – was invented (and, like the telescope, several opticians seem to have discovered it independently). A magnifying glass can show us details normally invisible to our eyes, and the seventeenth-century Dutch opticians made important new discoveries with these simple (one lens) microscopes. Anton van Leeuwenhoek, by profession a draper and the Delft City Hall janitor, ground precision lenses in his spare time. Keeping to small lenses for accuracy in grinding, he peered through pinhead-sized magnifying glasses at a tiny world which no one had seen before, discovering the carefully ordered structure of living tissue, and the existence of the one-celled creatures which we now know to be descendents of Earth's original life forms. Leeuwenhoek's lenses, magnifying almost two hundred times – enough to make a single letter fill a page of this book – reached the limits of magnifying-glass technology, and the way ahead was pioneered by the versatile English scientist Robert Hooke, who first put two lenses together in a tube to make today's type of *compound* (multi-lensed) microscope.

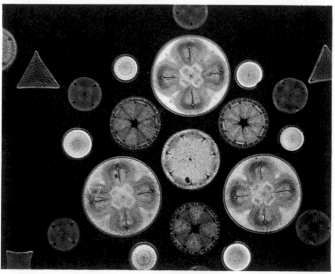

The onward rush of technology has long since perfected Hooke's compound microscope, and now calls for something more. For there's a natural limit to what even the best-made microscope can show us. Light travels as a wave motion, with a wavelength of less than a thousandth of a millimetre; and so light cannot show us anything smaller than this size. Think of ocean waves running into a rock. If the rock is a big one compared to the distance between wave crests – the wavelength – the waves cannot pass it in a straight line; their motion is changed. But when waves meet a small obstruction – perhaps the mast of a sunken ship, rather than a rock – their motion is practically unaffected. So it is with light waves. A comparatively large object, like a pinhead, affects the light waves which meet it, and the resulting disturbance is the image we see; but a single atom, or small molecule, is much too small to cause a ripple in the passing light waves.

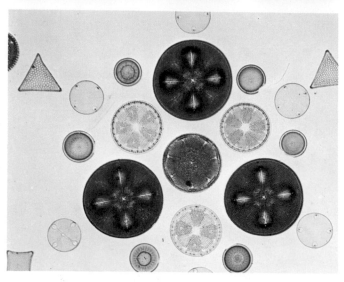

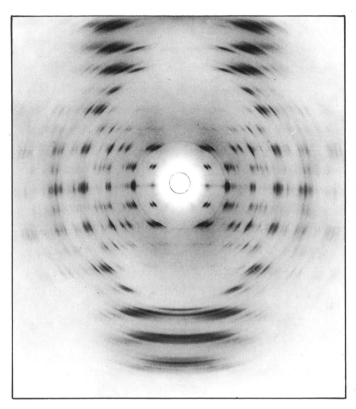

X-ray diffraction pattern of DNA; the X-rays emerge from the molecule more strongly in some directions than others. From photographs like this, taken by Maurice Wilkins of King's College London, the double helix structure of DNA was elucidated.

So we can't build a super-power microscope to look directly at atoms. Biologists still constantly use light microscopes to peer into living cells, for the basic architecture of the cell is visible, especially when the different structures are chemically stained. But there's no clue to the complex chemistry of life from this coarse overview. It's like looking at a city from the air: we see the general orderliness of streets and blocks of houses, but can't see the constant activity of humans which constitutes the life of the city.

Turn up the microscope's magnification beyond a thousand or so, and we reach Nature's limit. The details we see are as small as light's wavelength, slightly less than a thousandth of a millimetre. If we are looking at a gradually tapering wire, it will become less and less distinct as we approach this size, and the still narrower part will simply be invisible. We can't beat the limit by turning up the magnification: the images simply become bigger without losing their fuzziness – like enlarging a photograph which is out of focus – while anything smaller remains invisible. Higher powers just lead to what microscopists graphically call 'empty magnification'.

But we can draw on the fact that light is not the only wave in Nature's store. Shorter wavelengths of electromagnetic waves reach down to atom size and smaller, and those known as X-rays are the right wavelength to 'see' atoms as large obstacles. An X-ray microscope ought to show us the individual atoms in a piece of matter – but,

unfortunately, X-rays are notoriously difficult to focus. An ordinary microscope – or even a magnifying glass – only works because light rays are bent and focused by its lenses, and it is not easy to build a 'lens' for X-rays. As we'll see below, there are ways to achieve this sort of resolution (image-forming power), by using electrons. But X-rays are a vital tool for probing the way atoms are put together into molecules, and physicists have devised an ingenious lens—less system for this purpose. (It's in some ways similar to a laser recording a three-dimensional object on the flat photographic plate of a hologram without any lenses.)

In the science of X-ray crystallography, we start with a pure crystal of the substance, in which identical molecules are packed into a neat array. Beam the X-rays through, and some are bent sideways by the atoms in the molecules. The pattern of atoms, repeated over and over again in the molecular packing, ensures that the X-rays come out more strongly in some directions than others. Put a photographic plate in place to record the scattered X-rays, and it reveals a pattern of dots and lines. The skill of the X-ray crystallographer is in deciphering this intricate pattern, working backwards to uncover the layout of atoms. For simple molecules, experience and simple calculations are sufficient; but the molecules in living cells require more effort. The race to uncover the structure of the DNA molecule, which Francis Crick and James Watson won by a short lead, was largely a matter of sorting out the enigmatic traces left by X-rays which had been scattered by these molecules. Now such calculations have been handed over to computers, which can sort through possible configurations of atoms, and compare the predicted results with experiments in a fraction of the time a human researcher would take.

In an indirect way, X-ray crystallography shows up the individual atoms of matter. But German-American physicist Edwin Mueller has invented a completely different device, with which he can photograph individual atoms. In his field-emission microscope, an extremely fine 'needle tip' is centred in a vacuum cavity, and electrically charged. Helium atoms have been adsorbed on to the metal surface at an earlier stage and, as the needle becomes more positively charged, the helium atoms have their electrons drained away, and they become positively charged ions. These loosely held helium ions are now strongly repelled by the similarly charged needle tip, and they zoom off in straight lines. The very tip of the needle has a rounded shape (though on a very small scale), so when the electric field drives the helium ions out in straight lines, the ions falling on the top end of the vacuum chamber effectively project an image of the needle tip. The magnifying power of this 'microscope' just depends on how much bigger the chamber is than the needle tip. A million times magnification is quite possible, and at this scale images of the individual atoms

When light passes through a double slit, the two new expanding wavefronts 'interfere' to produce regions of intense disturbance (bright) and relative calm (darkness). Matter behaves similarly, but on a much smaller scale: disturbed – high energy – regions of 'matter waves' show where the matter particle is most likely to be.

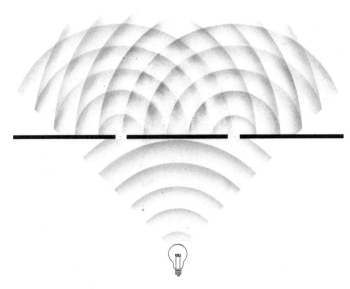

making up the needle tip can be picked out. Photographs of these enormously magnified images are already classics in science history, for they are obvious proof that matter really is made up of atoms. Since, for most people, seeing really is believing, one field-emission microscope picture is intuitively more convincing than the centuries-long efforts of philosophers, chemists and physicists to prove the atomic theory of matter.

But both X-ray crystallography and the field emission microscope are rather limited, despite their enormous magnifications. If we want a picture of the inside of a cell, as we would see it with a super-power ordinary microscope, we must turn to other devices. The electron microscope takes pictures using a surprising type of short-wavelength radiation: electrons themselves.

Wave-particles

We'll meet in the next chapter the nature of light, sometimes behaving like tiny particles, at others like an extended series of waves. And a few years after Max Planck and Albert Einstein had pointed out light's odd behaviour, an even stranger facet of Nature became apparent. Ordinary particles of matter – electrons, atoms, even golf balls and planets – can behave as waves.

This revolution was ushered in by a French nobleman, Prince Louis de Broglie. Struck by the way that physicists had already come to terms with light as both a particle and a wave, he discovered that the equations which describe how electrons move, or how a golf ball flies through the air, would work just as well if electrons and golf balls were packets of waves. These new waves are not of fluctuating electric and magnetic fields, like light and the other electromagnetic radiations. They are *matter waves*: 'a fluctuation in the position of matter itself' might be one way to describe

de Broglie's strange prediction in non-mathematical language. An understanding of how matter behaves as waves is essential when we look at the frontiers of research. To gain an insight into this, we can investigate how something commonplace, like a golf ball, is affected by its wavelike properties.

The easiest way is to begin with the wave-particle nature of light itself. Hold up two adjacent fingers almost touching one another, so that there's a gap of less than a millimetre (·039 inch) between them. Bring the gap up to your eye, look through it at a bright background like the sky, and you'll see thin, dark lines running the length of the gap. That's direct evidence that light is 'wave-like'. The light, disturbed by coming through the narrow gap, is subjected to 'interference', so that in some places all the 'peaks' of the waves meet up and make a bright line, while in others the 'peaks' coincide with the 'troughs' to cancel each other out and leave darkness. In terms of waves, the alternate bands of light and dark are easily explained.

But we can also think of light as travelling as a stream of particles. The sensation of seeing is just the impact of countless individual light photons on the backs of our eyeballs. The alternate light and dark stripes mean that photons mainly fall on to the particular stripes that we see as bright, and avoid the apparently dark areas. It's impossible to say where any individual photon will fall before it comes through the finger gap; but we can say it's most likely to fall in a 'bright band' region. This is how the 'wave' and 'particle' aspects of light tie together. Imagine a light photon travelling through the finger gap. If it were a simple moving particle – as we think of a rifle bullet, say – then it would move through with no diversion, and when we look at the light we would simply see an evenly lit strip. But the photon is wavelike, too. We can think of the wavelike part moving through the gap and interfering with itself to leave zones of no disturbance, and zones of great disturbance. Now the energy packet, or photon, is simply more likely to move in a direction which its associated wave makes a zone of greater disturbance.

There's nothing to say which bright zone the individual photon will end up in: it's just a matter of chance. When millions of photons are pouring in, however, it's clear that they'll end up fairly equally spread among the bright 'high disturbance' bands, while all of them will avoid the 'low disturbance' regions, leaving them dark.

If we think this all sounds like mumbo-jumbo, we are essentially right! The behaviour of light as both wave and particle makes complete sense when the relevant equations are worked out, but it's next to impossible to describe what's going on in everyday terms – because these aspects of light are just not everyday experience. But describing light as a photon whose motion is controlled by 'guiding'

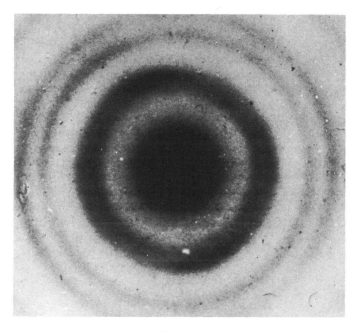

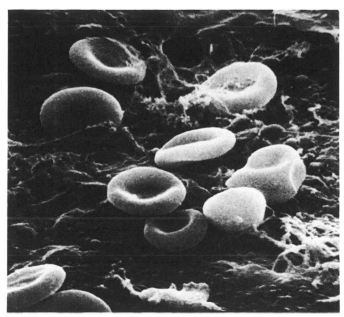

waves, which decide the direction it's most likely to move in, is a reasonable approximation.

And turning now to 'matter waves', we find – fortunately! – that we can think of them in a similar way. Any moving object has a 'guiding' wave associated with it, but in everyday life the wavelengths are so small that we just can't measure the effects. Take a similar example to light photons travelling through a gap. If we drive a succession of golf balls through an open doorway, theoretically its guiding wave interferes with itself to assert that there's a tiny chance that each ball won't go precisely straight through. Suppose that our aim is exact: without the doorway we could exactly hole the balls. Then, because of the 'wave nature of golf balls', the presence of the door means that each successive ball will reach the green in a slightly different place. To either side of the flag will be regions where golf balls tend to fall ('bright' on the light analogy), and regions where no golf balls fall ('dark'). The snag in practice is that these 'light' and 'dark' bands are separated by less than a million million millionth of an atom's diameter!

When we come down to the tiny masses of the fundamental particles, however, the guiding waves become very important. Indeed, the matter wavelengths can be inconveniently long. An electron travelling through the filament of a light bulb, for example, has a wavelength longer than the size of an atom. But speeding up the electron reduces its wavelength; and therein lies the secret of the electron microscope, the most versatile tool for looking at molecules and the structure of cells.

As well as having a wavelength much less than that of light (and similar to the electromagnetic wavelength of X-rays), a beam of electrons can easily be focused by magnetic fields. (Most of us have electronic 'lenses' in our own homes, in fact, for the magnetic coils around a television tube focus the electrons from the electron gun on to the screen.) The first electron microscopes just mimicked optical microscopes. A beam of electrons came up through the specimen, was magnified by an 'objective lens', and was cast on to a screen or a photographic plate. In more recent 'scanning' electron microscopes, the electron beam is bounced off the specimen's surface, to give three-dimensional-looking photographs.

Many of these scanning electron microscope pictures look more 'realistic' than conventional microscope photographs; it's hard to believe when we see them that we're looking at a world a hundred thousand times smaller than real life – where a human hair would stretch 10 kilometres (6·2 miles) and overtop a four-storey house in width. This is the limit to which we can get detailed pictures, as yet; to probe further, to within the atom, needs more roundabout techniques.

Inside the atom

Ernest Rutherford, the New Zealand physicist who first transmuted one element into another, was also the first man to peer inside the atom. His method foreshadowed many later experiments in tracking down the very small, and it was essentially very simple: just fire some kind of particle at a target, and see what happens to it. His projectiles were the high-speed helium nuclei naturally thrown out by radioactive elements, and the target thin gold foil. He predicted a straightforward result, for at that time scientists thought that atoms were diffuse, spread-out blobs (with electrons embedded in them), and the high-speed particles should have gone straight through. And although most did, a few came straight back. As Rutherford put it: 'It was almost as incredible as if you fired a 15-inch shell at a piece of tissue paper and it came back and hit you.'

Only one explanation fitted. Almost all the atom's mass, and all its positive charge, must be concentrated at the centre in what Rutherford termed the nucleus. Practically the entire volume of the atom around it is empty space, containing just the orbiting electrons, moving around the nucleus. Today this picture of the atom seems very familiar; but in 1910 it was a revolutionary step forward in our understanding of the Universe.

Nearly sixty years later, such 'target practice' experiments started revealing what's inside some of the supposedly 'fundamental' particles. Shoot an electron at another electron and they bounce apart like a pair of billiard balls. Crank up the first electron to a higher speed to begin with, and it has a shorter matter wavelength: the two particles bounce apart like smaller balls. Even at the highest speeds which physicists give electrons, they always bounce apart as simple spheres, whose effective size goes down with increased particle speed, simply because the wavelength decreases. Physicists think that electrons really are simple, point-like particles.

But when electrons bounce off protons, the picture is entirely different. Electrons with short enough wavelengths are reflected off in an unusual way, showing that protons are not simple points of matter. Instead, inside each proton there seem to be three points where mass is concentrated. At first these were called 'partons', but now it's obvious that they have exactly the same characteristics which theorists had predicted for the three quarks making up a proton. These more recent scattering experiments, using electrons whose enormous speeds give them wavelengths small enough to 'get inside' the protons, are showing us quarks themselves. And indeed, as physicists have not yet succeeded in getting a 'free' quark in isolation, the electron (and muon) scattering experiments are the closest that scientists have yet come to 'seeing' individual quarks.

At this point, we should stand back and realize that on the very small scales we are talking about in particle physics, the notion of *size* is rather ambiguous. An electron probably has no real 'diameter'; it is infinitesimally small, a literally zero-sized point of mass and electric charge. But in an actual experiment, the electron behaves as though it has a size roughly equal to its wavelength: increase its speed, and its effective size goes down.

A slow proton behaves as a particle the size of its wavelength too, but at higher speeds its decreasing size does reach a limit. For a proton does have a meaningful real size, which is roughly the distance separating the quarks which make it up.

As we noted before, the world of the very small makes few concessions to common sense. Particle physicists don't generally talk in terms of 'sizes', for measured sizes generally depend on how the experiment is done. And even masses are sometimes difficult to define. Quarks are bound tightly in a proton, and the energy of binding – which is equivalent to mass – means that the sum total of the three quark masses is not the same as the proton's mass. A particle physicist's language abounds in exotic terms, some fairly familiar, like energy and electric charge; and others which are far from everyday experience: strangeness, charm, isotopic spin, parity, and baryon and lepton numbers. Yet physicists have found that they need a plethora of new terms to describe the uncommon properties and reactions of the fundamental particles which make up our Universe; on the very small scale it makes sense to use terms like these rather than everyday concepts like size.

And at this level, experimental results are enmeshed with theory. We can't see individual quarks or leptons, and the evidence for them comes indirectly, from flashes of light or surges of current in particle detectors, and from curved tracks on photographic plates. Many of the most exotic new particles don't even show up in these ways, but physicists must invoke them to explain the production of known particles, in unexpected combinations, paths and energies. All this may seem like fumbling in the dark; but results of different experiments interpreted by different research groups around the world do hang together well, and the vast majority of experiments do confirm the predictions of current theory. Only a few come up with unexpected results, and these generally extend the theory – by bringing in new types of particle, for example – rather than contradicting it. There's every reason to believe that the odd world of quarks and leptons, with their peculiar properties and exotic reactions, is as real as the larger-scale realm of atoms – and, after all, the existence of atoms and molecules was for a long time based only on roundabout experimental evidence.

Pursuing the unseeable

The 1960s saw the 'space race' to put a man on the Moon. After great expense and the development of the new technology of giant rockets and space travel systems, the Americans 'won' when Neil Armstrong set foot on our sister world in 1969. The 1970s and '80s are witnessing a new, less heralded race – and yet one whose prizes are greater than national prestige or the returning of a bagful of Moon rocks. In huge laboratories in the United States and Europe, physicists are chasing the ultimate particles of which the Universe is made. And the prizes are not just the discovery of new particles. Each new experiment fits in another piece of the complex jigsaw of fundamental particles and forces, and one day enough of the puzzle should be fitted together to enable us to stand back and see the overall pattern of the basic make-up of our Universe.

The laboratories must be huge, for they involve machines which accelerate particles up to enormous speeds. In principle it's simple. Put an electrically charged particle in an electric field, and it is pulled by the electric force. The electrons in a television tube are speeded up in exactly the same way, dragged through the vacuum in the tube towards the positively charged screen by the electric force. When they impact on the phosphor screen, their energy of motion changes into the flashes of light which build up to produce the picture.

But particle physicists need higher-speed electrons than this. At Stanford, California, electrons are continuously accelerated along a vacuum tube 3 kilometres (nearly 2 miles) long. The elongated mound covering this sophisticated machine stretches over the countryside like an Iron Age earthwork, passing under highways when it meets them. At the end, these electrons shoot out into a whole barrage of experiments, to collide with other particles and yield up some of the answers to the particle physicists' questions. Many particle accelerators are built in the form of large rings, in which the fast particles travel round and round, kept to the centre of the vacuum tube by magnets all along the ring. The largest European accelerator, at the CERN laboratory near Geneva, is a recently completed ring 7 kilometres ($4\frac{1}{3}$ miles) in circumference, and it accelerates protons to 99·9997% of the speed of light. And the cost of these installations reflects their size and complexity: the new European accelerator, for example, set back the participating nations 1,100 million Swiss francs (250 million dollars at 1971 prices).

It seems ironic that physicists need such huge and expensive machines to study the smallest things in the Universe. Yet that's the way it works out. We've met part of the reason already: the wave nature of matter. To get 'inside' and see how particles like protons are made of quarks, we need to probe them with very short-wavelength, high-speed particles.

But the most exciting results from particle accelerators are the new particles: in 1974 the first particle to contain charmed quarks; and later, a particle probably containing a new ('bottom' or 'beauty') quark; while over the same period, evidence has accumulated for a fifth member of the lepton family. As we saw in Chapter 3, it's impossible to produce a new particle out of nothing; but if we concentrate enough energy in one place, we may get a particle and its antiparticle together. Smashing a high-speed particle into another is an ideal way of getting such an energy concentration, and the collision debris is usually a rich mixture of particles and antiparticles. And just occasionally,

a previously unknown pair will turn up.

The simplest way to provoke such a collision is just to let a beam of high-speed protons, electrons or whatever fall on to a fixed target. The target, perhaps a block of iron or a chamber filled with liquid hydrogen, can be quite thick, so that there's a good chance that the beam particles will collide with a nucleus within the target, to produce an 'event'. The tracks of the collision debris from these events tell physicists just what kinds of particles (and antiparticles) have been produced. And because so many events happen in 'fixed target' experiments, particle physicists can measure up a fair number of the comparatively rare events which produce new particles. Taking the average of these results gives the experimenters quite accurate values for important things like the masses and lifetimes of the new particles.

Fixed-target experiments, however, don't make full use of the incoming particles' immense energies. Imagine that they are racing cars speeding around a circular track. Our target 'nuclei' are other cars parked with their brakes off in a side road. To start the experiment, the racing cars are diverted into the side road, and they collide with the stationary cars, with a splash of debris. But much of the racing cars' energy will be used up in smashing into the stationary cars, pushing them forward so that their new motion carries away some of the energy which would otherwise end up in the debris.

There's a fairly simple way around this problem. We build a second racing track, roughly the same size as the first, but with a slightly different shape so that the two tracks intersect at only two places. Now we send the racing cars around one track clockwise, and around the other anticlockwise. At the intersection points, there's a chance that a car from one track will collide head-on with one from the other track coming the other way. If both cars have the same mass, they will stop dead, and all the energy of motion ends up as collision debris. The price we have to pay for these much higher-energy collisions is that they are comparatively rare: there's a much smaller chance that a racing car will meet one coming the other way at an intersection than that it will collide with one of a crowd of stationary cars in the fixed-target experiment.

But these high-energy collisions are important when we are looking for new particles. According to Einstein's mass-energy formula, the higher the energy present at the collision, the more massive the type of particles and antiparticles that can be made. Experimenters found the new lepton with the intersecting storage rings at Stanford, simply because it could produce high-energy collisions. Electrons and their antiparticles, positrons, from the long accelerator there were 'stored' in two intersecting rings, by keeping them whirling around in opposite directions. At

In an electron microscope, electrons bounce off the outer layers of atoms; the field emission microscope uses electrically repelled helium nuclei to image atoms themselves. The internal structure of atoms

was revealed by firing high energy helium nuclei through them, and noticing that some came straight back; high energy electron beams can now show the quarks making up the protons and neutrons of the nucleus.

the intersections, the high-speed electrons and positrons could collide and annihilate each other in a flash of energy so intense that particles and antiparticles 7,000 times heavier than the electron could arise in the debris, as energy spontaneously turned into matter-antimatter pairs.

Even more powerful electron-positron rings are now being built at Stanford and in Germany, to try to create even more massive particles. Proton-proton and proton-antiproton colliding rings are also being planned in both Europe and the United States, as particle physicists vie in the race to find yet heavier new particles. And it isn't just a question of particle 'collecting', as some people collect stamps or coins, for each new kind of quark or lepton is telling us something of how the Universe is made up. Number one priority, however, is the search for heavy particles which are neither quark nor lepton: the 'W' and 'Z' particles, which are particle-like manifestations of the weak nuclear force itself.

Virtual energy holds it all together

To understand what this latest goal of particle physicists actually *is*, we must again hold tight and plunge into the odd ways that Nature behaves on the small scale. So far we have talked of two families of particles – the quarks and the leptons – and the four forces of Nature. But what are these forces? How can two particles separated from one another still influence each other? Physicists have long distrusted the idea of 'action at a distance': they prefer to think in terms of some tangible connection between the two particles.

To see what this connection could be, let us return to the rather more familiar electromagnetic force. Two negatively charged electrons approach; their like charges repel; and they push themselves apart under the influence of their electric fields. That's the action-at-a-distance way of looking at it. The quantum theory shows, however, that there is a tangible link between the two electrons: a very fleeting exchange of photons. Photons (next chapter) are the energy packets that light travels in; they are effectively a kind of particle, but in neither the quark nor the lepton camp.

In one sense it's not surprising that photons should be involved, for they are particles of electromagnetic *radiation*, and the electron-electron repulsion is due to the electromagnetic force. But when we start to ask where they come from and go to, the theory gives a strange answer: they never actually exist!

The roots of this odd situation go back to the wave-particle nature of all matter. The German physicist Werner Heisenberg pointed out in 1927 that de Broglie's wavelike matter had a consequence of shattering philosophical, as well as scientific, importance. He called this the *Uncertainty*

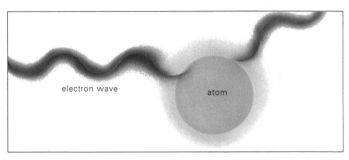

electron wave   atom

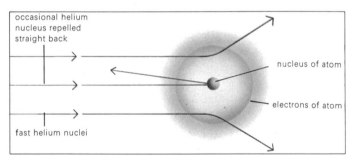

photographic plate

highly-charged needle tip

helium nuclei

images of individual atoms

atoms

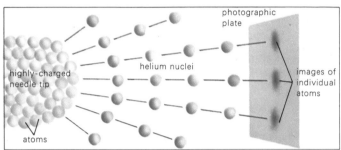

occasional helium nucleus repelled straight back

nucleus of atom

electrons of atom

fast helium nuclei

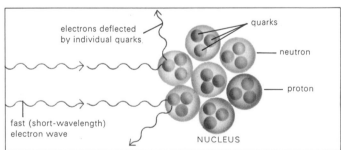

electrons deflected by individual quarks

quarks

neutron

proton

fast (short-wavelength) electron wave

NUCLEUS

Principle. There is a natural limit to just how much we can know about the Universe. However greatly physicists refine their measurements, eventually they will come to the point where Nature's own limit prevents them from finding more.

To see how this works, let's take another look at the not-too-serious example of an ace golfer driving through a doorway to a distant green. If the doorway weren't there, he would hole exactly. Or, to put it another way, the golf ball would not deviate sideways at all from a straight line towards the hole. But (neglecting light, for the moment) there would be no way of knowing the ball's position on its way to the hole. Here's a justification for our doorway: if the ball reaches the green, it must have passed between the

The large particle accelerator
at the Fermi National Accelerator
Laboratory, Illinois, is 6·5 km
(4 miles) in circumference. The
tremendously energetic protons
in the ring are used to probe
the quark structure of particles,
and to create pairs of new
particles and antiparticles.

doorposts, and so we know its position. And here is also the problem, for the wave nature of golf balls means that it may be deflected slightly from its straight course. By defining its position with the doorway, we can no longer be certain of its exact direction of travel.

Such problems do not worry the practical golfer, of course. He can see the ball at all times (though the light photons reflecting off the ball actually make its trajectory much more uncertain than the fundamental minuscule deviation due to Heisenberg's Uncertainty Principle). But if we replace the golf ball with an electron, the doorway with a narrow slit, and the green with a photographic plate, the consequences are obvious. The more we try to define the electron's sideways position by narrowing the slit, the more its wave nature will make its ultimate direction uncertain.

This all fits in with Nature's forces in a logical, though apparently roundabout way. The Uncertainty Principle also shows us that we cannot measure energies accurately unless we can make the measurement over a period of time. For everyday energies, the time we need is very short, and there's no problem. But when we're dealing with the much smaller energies which the fundamental particles have, the times needed are longer. Let's turn this into a more useful way of thinking about forces: it is quite possible for a packet of energy to spring spontaneously into existence in a vacuum, as long as it disappears again in a short time – so

short that Nature's limit prevents us from measuring it. In a way, it's Nature's version of 'what the eye doesn't see, the heart doesn't grieve about': if the energy exists for too short a time to measure, it isn't actually breaking the law of conservation of energy (which says that it's impossible to create energy from nothing).

So, when two electrons are seemingly repelling each other by the electric force, they are actually exchanging photons which exist too fleetingly for us to measure directly. This may not seem any better a description than talking of 'force fields', but physicists can do calculations based on this exchange of 'virtual' photons, and the results agree extremely well with many kinds of experiment. This is the theory of *quantum electrodynamics*, and it is one of the most exact in physics today; it ties together all the phenomena of electric and magnetic forces with other properties – such as spin – of the fundamental particles, and accounts for all the wave and particle aspects of light and other electromagnetic radiation.

So physicists see the photon as a particle with two roles. A straightforward photon is just radiation, like light, travelling from one place to another. But virtual photons, existing for too short a time to measure, are the carriers of the electromagnetic force. And naturally, physicists have deduced that the other three forces must be carried by virtual particles. These can't – by definition – be detected in

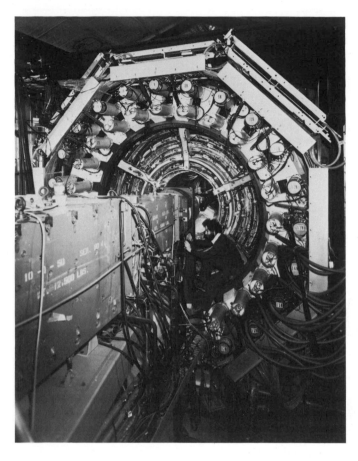

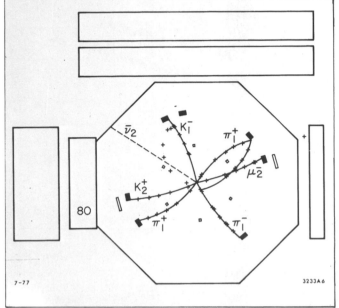

their role as a force carrier; but, like photons, they should also exist in a 'real' form, travelling as a new type of radiation.

According to Steven Weinberg and Abdus Salam's theory, which combines weak and electromagnetic forces, the photon is only one of a family of four. It differs from its brothers in having no mass, and this is the reason why electric and magnetic forces operate over very long distances. The other three, weak-force, particles (one called 'Z', and two called 'W') are heavy, probably some 70 times heavier than the proton, and theory shows that this will restrict them to the very short ranges which we know the weak force has.

So the current race among particle physicists is for the W and Z particles, and they should turn up in the higher-energy accelerators now being planned in Europe and the United States. They won't be just new particles to add to the collection, for their properties will tell physicists just how far they are right in the combination of electromagnetic and weak forces, our present step in the unification of Nature's forces.

Strong glue for the quarks

The strong force, too, is mediated by hypothetical particles,

aptly but inelegantly called gluons. Gluons bind the quarks together in threes to make the baryon particles, such as the proton or neutron; and they can also hold quark-antiquark pairs together as mesons. Free gluons are not likely to turn up in present particle physics experiments, but physicists have been able to learn a lot about quarks and their associated gluons from the properties of these tightly bound quark groups, the old 'fundamental particles'.

The strong force is indeed very strong. So far, at least, it's proved impossible to knock a quark out of a baryon against this restraining force. There was a flurry of excitement in 1977, when it seemed that physicists had detected 'free' quarks in small metal balls (on the basis of their having an electric charge less than an electron); but later, more sensitive experiments haven't confirmed their results. Indeed, there are theoretical reasons for believing that quarks cannot exist in isolation.

Inside a proton – or other baryon – the quarks seem to lead fairly independent existences. We have already noted how high-speed electrons can pinpoint three 'centres of mass' inside the proton. The strong force seems to act like a piece of elastic: when two quarks are close together, they affect each other very little; but try to pull them apart, and the force becomes more and more powerful. (All the other forces of Nature act the other way around: the closer two particles are, the stronger are the electromagnetic, weak or gravitational forces.) This unusual behaviour of the strong force actually fits in very well with the theory of how quarks stay together in threes by exchanging 'virtual gluons'. The theory, called 'quantum chromodynamics', predicts a family of eight different gluons involving themselves in the strong force.

Probing the baryons' structure with fast electrons not only shows the three quarks, but also reveals that the gluons make up about half the mass. So the strong force, as well as binding the quarks together, actually contributes something of its own to the resulting particle. A proton is as much 'glue' as quarks.

In other experiments, particle physicists have investigated the inner workings of protons by smashing them together. In the burst of energy liberated, 'jets' of new particles and antiparticles shoot off in opposite directions; and by studying these jets, physicists can say something about the quarks within the original protons. Again, the results confirm gratifyingly well the theorists' predictions for the electric charges and spins of the quarks.

So the quark theory is alive and well. Unfortunately, the present theory doesn't predict just how many different kinds of quark there should be. In the original theory, three were needed – up, down and strange – while the fourth, charm, was soon postulated. This was actually 'discovered' in 1974, almost simultaneously in two different experiments, which earned Nobel prizes for the team leaders, Burton Richter and Samuel Ting. Their findings, announced on 11 November, are fondly recalled by particle physicists as the 'November Revolution', for they were not just a triumph for quark theory, but they indirectly bolstered the joint electromagnetic-weak force theory.

The 'discovery' of the charmed quark doesn't mean that a single isolated quark turned up in the experiment. The particle found was actually a combination of a charmed quark and a charmed antiquark, as a type of meson. It couldn't be explained as a combination of the previously known quarks, though, and further research has proved the initial interpretation right.

And in 1977, the story was repeated, when American physicist Leon Lederman discovered a new meson which could not be made from any of the four known quarks: it must be a new quark-antiquark pair, which theorists had already named either 'bottom' or 'beauty'. Theory suggests that the number of quarks should be even, and now the search is under way for the sixth type of quark: 'top', or more elegantly, 'truth'.

And so to the last of Nature's forces, the extremely weak and enigmatic gravitation. Particle physicists can't experiment with gravitation, for it just doesn't have a measurable effect on the fundamental particles in the laboratory. But theoretically, it, too, should be the result of virtual-particle exchange, involving a particle dubbed the 'graviton'. Gravitation is something of an outsider among the forces in two ways. It always attracts, and never repels; and the strength of the force depends on a body's most fundamental property: its mass. Never mind whether a particle is electrically charged, or whether it's a quark or a lepton – gravitation will affect it the same way. Albert Einstein took gravitation to be a special force, regarding it as a distortion of space itself; and his theory is still the most useful one to astronomers, as we'll see in later chapters. But gravitation can be described as an exchange of virtual gravitons, too.

What makes the graviton different from the other 'force particles' – the proton, W and Z particles, and gluons – is its *spin*. We've already mentioned that Nature has a basic unit of spin. Particles like quarks and leptons have half a unit each. All force particles apart from the graviton have one unit of spin; that is, in some sense, they have twice the quantity of spin around their own axes. When quarks or leptons exchange these 'force particles', the theory shows that the spin of one unit means that 'like' particles (quarks and leptons) will repel each other – as we know happens in practice. But each graviton has *two* spin units; and such particles will always cause a force of attraction – and with a strength depending on the body's mass. The particle physicists' theory can explain the unusual properties of the gravitational force simply by the high spin rate of the virtual particles (gravitons) which are exchanged.

We may think that this is putting a lot of emphasis on something as trivial as spin. Yet this quantity is a surprisingly powerful influence in fundamental physics. Particles with half a spin unit (quarks and leptons) can never exist in close proximity to another such particle with exactly the same properties. That's why electrons orbiting the nuclei of atoms don't all pile up in the innermost orbit. Once that is filled (by two electrons with opposing spins), the next electron must go in a further-out orbit. So arises the complex outer structure of the atoms, which produces all their different chemical properties.

But particles with 'spin-one' or 'spin-two' prefer to bunch together. From this point of view, a laser beam is a bunch of exactly identical photons, all with exactly the same energy, and liking it that way.

So the physicists' world is populated with half-spin particles of the quark and lepton families; and the whole-number spin particles, which are a law unto themselves. Exchanged between the quarks and leptons as quick, unseeable 'virtual' particles, they cause the forces of Nature. At least one, the photon, also exists as a particle in its own right, and with powerful enough particle accelerators the gluons and W and Z particles should begin to reveal themselves.

These particles, and the new quarks and leptons, will help physicists to unravel the ultimate secrets of the Universe, to sort out which ingredients are really basic, and which are perhaps just of secondary importance. The new particle accelerators are a step on the road to this goal; and, if past experience in science is anything to go by, they'll reveal more than these predicted particles.

The glowing curtains of the
aurorae – Northern and Southern
Lights – are caused by electrically
charged particles from the Sun

striking the Earth's upper
atmosphere, following the Earth's
magnetic field inwards.

The night sky is a black canopy speckled with thousands of tiny, sparkling jewels. And to an astronomer, starlight is carrying information from the depths of space.

The very existence of lights in the sky has always fascinated Man, and it's no coincidence that astronomy is the oldest of the sciences. The movement from week to week of some of the brighter lights – planets, which means 'wanderers' in Greek – eventually led to an understanding of our solar system; and the recognition of other stars as distant Suns gave men some comprehension of the immensity of the Universe. Deciphering the message of starlight is the prime concern of today's astronomers, because, apart from space probes to the planets, virtually everything we know about the Universe has to be worked out from the light and other electromagnetic radiations that we receive on Earth.

The invention of the telescope brought astronomy out of its Stone Age. It's not certain who actually invented the first two-lens telescope: the principle was probably discovered independently by several opticians in the late sixteenth century, when the first lenses were made that were good enough to give clear images when arranged as a telescope. There's no doubt, however, that the famous Italian scientist Galileo was the first man to make extensive astronomical observations with the new instrument. And the results were astonishing: round craters on the Moon; small moons circling Jupiter; the planets revealed as globes rather than just points of light.

Galileo's first telescope magnified some thirty times, but probably showed no more detail than a good pair of modern binoculars with much superior lenses. Everyone knows that a telescope magnifies the apparent size of distant objects; but that isn't the modern astronomer's main concern. There's a very real problem facing astronomers at the telescope: the 'twinkling'. This constant shimmering of stars is due entirely to the Earth's atmosphere, which is in constant motion. The light coming down from above is constantly bent and shifted, and so the image of a star seen under high magnification at the telescope is continually wandering and jumping about. When looking at a planet, the reflected light from each bit of the surface is shifting, and the image is continuously distorted. It's like looking at the bottom of a swimming pool from above, when the water is being disturbed by swimmers. Just occasionally, there are short moments when one is looking through still air, and the trained observer has to memorize as much detail as possible in that rare and magical moment of 'good seeing'. Yet even so, a magnification of a few hundred times is all that an average night's seeing will allow.

In fact, professional astronomers no longer look through telescopes these days, for the human eye is not a very good observing instrument. Photographs have several great advantages, and a telescope in a professional observatory is now used almost exclusively as an astronomical camera. Indeed, a telescope used in this way is identical in optical layout to a photographer's long-focus or telephoto lens, the only differences being that of size, and the fact that large modern telescopes use a mirror, rather than a lens, to collect light.

Photographic plates store up all the light that falls on them, and so, unlike the eye, they can record fainter objects by means of long exposures. And a plate is a permanent record, which astronomers can study at leisure. One of the eye's most serious drawbacks is the subjective distortion which creeps in when we strain right at the limit of visibility: the 'canals' of Mars are perhaps the prime example of the eye misleading the brain.

But if magnification is not particularly important, why are astronomers building huge telescopes, up to the size of the 5-metre (200-inch) diameter telescope at Mount Palomar and the Russian 6-metre (236-inch) at the Zelenchukskaya Astrophysical Observatory? The answer is simply that a larger telescope collects more light than a small one. Many of the most exciting objects in modern astronomy appear faint, simply because they are so distant, and large telescopes must be used to collect enough light to study them.

A photograph taken with a large telescope actually shows less detail on bright objects than the eye can see at the same

telescope. Exposure times are typically several minutes, and in this time the moments of good seeing are swamped by the normal shifting patterns which blur out the fine detail. Future prospects are better: a small, flexible mirror which distorts to counteract the atmosphere's distortion has already been tested, and gives clearer images. Even stranger is the technique of 'speckle interferometry', where the actual pattern of atmospheric distortion of a star is photographed with a very short exposure, using an electronic intensifier. By processing a series of these pictures in a computer, it's possible to remove the distortions. Very close double stars can be 'resolved' in this way; and some of the nearest giant stars appear not as points of light, but as recognizable discs.

The major breakthrough, however, will come with the Space Telescope, which will be put into orbit in the 1980s. Above Earth's distorting atmosphere, this telescope will be

able to photograph details ten times smaller than Earth-based telescopes can see; and it can see fainter objects too, for it doesn't have to contend with the background of light from the atmosphere, which eventually fogs ground-based astronomical photographs during long exposures.

Waves of colour

The French positivist philosopher Auguste Comte declared in 1835 that there are certain facts which science can never ascertain, and as an example he chose the constitution of the stars. Certainly, it is not evident that, even with large telescopes and sensitive detectors, we can learn much about the composition of the stars. But Comte's example turned out to be a bad one, for since his time, scientists have delved far more deeply into the miracles of light itself. Unknown to Comte, the light from a star tells us not only where it is, but also its composition, its temperature, the strength of its gravitation, how rapidly the gases at its surface are moving about, and how fast the star is moving towards or away from us. To understand this, we must look at the nature of this subtle 'celestial messenger' (to quote the title of Galileo's most famous work).

Light travels as a wave motion. The English scientist Thomas 'Phenomenon' Young, whose other accomplishments included tightrope walking and the first translation of some of the hieroglyphic symbols on the Rosetta Stone, proved in 1801 that light rays can sometimes cancel each other, leaving darkness. The only way to explain this problem was to assume that light is some kind of travelling wave, so that sometimes the 'peaks' of one light ray can coincide with the 'troughs' of another. The net effect is a cancellation of the wave motion, just as ocean waves approaching from different directions around a headland can leave small patches of undisturbed water where they meet.

The wave theory was put on a firm footing by Scottish physicist James Clerk Maxwell, who showed that the 'undulating' nature of light was simply due to varying electric and magnetic fields. He was combining all the observations of electric and magnetic phenomena into one all-embracing electromagnetic theory, when he realized that the equations predicted a type of radiation consisting of moving electric and magnetic fields. The velocity that he calculated for this was exactly the same as the already measured speed of light, and further calculations left no room for doubt: light is electromagnetic radiation. For example, we all know that light rays are 'bent' – refracted – when they pass from one transparent substance into another, as from air into a lens or a glass of water. From Maxwell's theory this result is entirely expected: the electrons in the solid interact with the undulating electric field of the light ray, and change its direction.

If light is a wave, it must have a *wavelength*. The wavelength of an ocean wave is the distance between one crest and the next; for light, it is the distance between one peak in the electric field and the next – not as easy to visualize, but it does exist. The radiation that we call light actually covers a range in wavelengths; and in fact, we are analyzing the wavelengths of light every day of our lives, for the sensation of colour is caused by the different wavelengths. Red light is the longest wavelength we can see, with a difference between successive crests of 700 millionths of a millimetre (700 nanometres). Slightly shorter wavelengths, about 560 nanometres, are seen as yellow; those around 500 as green; while the shortest we can see are blue and violet (about 420 and 400 nanometres respectively). Intermediate wavelengths are seen as intermediate colours, such as orange and blue-green.

This sequence of colours, red-orange-yellow-green-blue-violet, is very familiar: it is the order of the colours of the rainbow. The raindrops in the air can split up the colours in 'white' light (a mixture of all visible wavelengths) and spread them out as a spectacular bow across the sky. At any time when light is refracted on moving into a different transparent substance, the wavelengths are bent to slightly different degrees, and are spread out into a panoply of colours.

So we can study the light from any object, light bulb or star, by passing it through a suitable transparent block. The simplest is a triangular glass prism, and the spread of colours on the far side is the *spectrum* of the shining source.

Although a prism is the simplest means of making a spectrum, some light is absorbed in the glass. Astronomers are usually concerned with faint objects, and they prefer to split light with the less light-wasting *diffraction grating*. Extremely fine parallel lines ruled on the surface of the grating, about 500 to every millimetre (0·039 inch), act as tiny reflectors. The reflected rays of light find that they

cancel one another out – peak for trough – in all directions except one. And this non-cancelling direction depends on the wavelength of the light, so white light is again spread out by the grating into the spectrum of colours. (Incidentally, pieces of 'jewellery' and decorative badges with crude grating surfaces are now made this way – they seem to be just flat metal, but reflect different colours as you turn them.)

Let's look closely at the spectrum of a typical star, the Sun. In working out its message, we have to allow for the fact that our eyes are most sensitive to yellow light, and fall off in sensitivity towards the longer red wavelengths and the shorter blue. Even so, accurate measurements show that the Sun's spectrum is brightest around the yellow-green region, about 520 nanometres. Not all stars follow this pattern. The red Betelgeuse in the constellation Orion is brightest at wavelengths beyond the red end of the spectrum, in the very long wavelength region called the infrared. Its visible light is, not surprisingly, concentrated towards the red end, and so this is the colour it appears to shine. The other bright star in Orion, Rigel, has its spectral

peak in the very short-wave ultra-violet region – the invisible radiation beyond the violet end of the spectrum – and its visible light is predominantly blue – violet being a dim colour to our eyes. All the stars shine in all wavelengths of visible light, though, and the coloured appearance that we see – reddish Betelgeuse, blue Rigel – depends on how the radiant energy is distributed through the spectrum.

We can easily understand the colours of stars as a consequence of their temperatures, as we saw in Chapter 2. A poker in the fire glows red-hot, but in a blacksmith's forge it would shine yellow-hot, while the even hotter filament of a light bulb is at yellow-white temperatures. Even though we think of both light bulbs and the Sun as shining yellow-white, the former are actually at a far lower temperature, and are more orange-coloured as a consequence: compare the colours of a light bulb and the Sun directly, or take a colour picture under artificial light, and the difference is quite surprising.

Physicists have long studied objects which glow simply because they are hot, and they use the seemingly perverse name 'black bodies' for them. A physicist's black body is actually an ideal. It is a body which absorbs all the radiation which falls on it, without reflecting any. A sooty chimney is the closest that most of us come to meeting a black body, but even this isn't a perfect absorber of light. Physicists think in terms of the inside of a large box with a tiny hole: any light which gets in through the hole is likely to be absorbed by the inside walls before it is reflected enough times to find its way out again through the hole. If the box is heated up, its inside will begin to glow; and then the radiation it emits through the hole can only depend on its temperature, and not on radiation falling into it from the outside.

Physicists measuring the light radiation from such boxes found that these glowing light sources emit all wavelengths of light, but they radiate most intensely around a certain part of the spectrum – just as astronomers found for stars. The wavelength of this 'peak' of emission gets shorter for hotter and hotter bodies. So these laboratory studies linked the wavelength of maximum light with measured temperatures; and astronomers now routinely 'take' the stars' temperatures by comparing the strength of the spectrum at two different wavelengths (usually yellow-green and the shorter blue), interpreting the results in terms of the physicists' black-body radiation.

Energy in packets

At the turn of this century, astronomers could say why some stars are red, others yellow and some blue, in terms of their different temperatures. But theoretical physicists were in confusion over the results of the black-body experiments. From the apparently straightforward observation

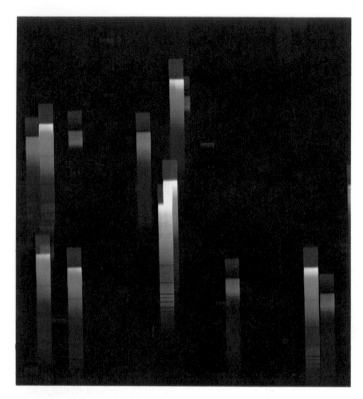

that a 'warm' body glows red, and a hotter one yellow, was to spring a revolution which knocked aside some of the foundations of their established physics. And this new *quantum theory* was just as important as the more famous relativity theory in establishing the modern scientists' outlook on the Universe.

The eminent Victorian scientist Lord Rayleigh first worked out the theory of light emission from black bodies. Drawing on all the basic axioms of the physics of his time, he unambiguously calculated the result – and his answer disagreed almost totally with the results of experiment, and with everyday experience. Rayleigh used his calculations to 'prove' that however hot or cold a body is, it will emit most of its light at very short wavelengths, much shorter than those of visible light; it was, apparently, quite impossible to have a spectrum whose intensity peaked at visible wavelengths. Clearly the theory was wrong; and yet Rayleigh had based his calculations on the firm foundations of established scientific laws.

The German physicist Max Planck determined to solve this problem, christened 'the ultra-violet catastrophe'. It boiled down to the fact that in 'classical physics' every long wavelength can produce effectively infinite numbers of shorter wavelengths in association. Students of music know that a low note on the piano gives rise to a whole series of harmonics of higher pitch (shorter wavelength); and the case of light should be somewhat similar. Planck had to find how Nature limits the number of short wavelengths

which a black body actually produces.

He came up with a bold suggestion, but one which did not clash too violently with the well-established laws of physics. Suppose that at the point where the light wave was emitted, the heat energy had to exceed some particular threshold. And that each wavelength had a different threshold, with the shorter wavelengths needing a higher energy before they could be radiated. Now the black body's surface can quite easily produce long wavelengths, for which only low energies are necessary. As we go to slightly shorter wavelengths, the intensity of the radiation will begin to increase, according to Lord Rayleigh's law. But as we look at still shorter wavelengths, we get to the point where the very short waves need a very high concentration of energy before they can be emitted at all, according to Planck's new principle. The black body finds it difficult to emit these short wavelengths, and so the intensity begins to fall off again at the short-wavelength end of the spectrum. There must be a peak somewhere in the spectrum whose wavelength depends on the black body's temperature. Planck's theory agreed excellently with the experiments, and physics seemed saved without having to sacrifice any of its basic principles.

But even Planck did not realize the full enormity of the revolution which he had begun by associating an energy with the radiation of a particular wavelength. He thought it was something peculiar to the way a hot body emits a wave of light. It was Albert Einstein who had the insight to realize that certain experiments in which light could push out electrons from metal surfaces extended Planck's theory into a veritable revolution. He pointed out that when a wavelength was emitted by a certain threshold concentration of energy in the black body, it carries that precise quantity of energy with it. On hitting a metal surface, the light wave gives up precisely this amount of energy, and can eject an electron if circumstances are favourable. So light is not just a wave. Light travels as individual packets of energy. The word *quantum* (Latin for 'how much?') was coined for an energy packet, and the Planck–Einstein revolution emerged in the form of the quantum theory.

A packet of energy which stays together as it flies through space is essentially a new kind of *particle*, and the quantum of light is called a *photon*. The quantum theory is a whole new way of looking at the world, for it demands that light should be both a particle and a wave. Sometimes its energy appears concentrated at a point, as only a small particle could achieve; while at other times light can cancel out with light to give dark fringes, an undoubtedly wave-like effect. But this confusion over the 'wave-particle' duality of light is really in our own minds. It only arises when we try to think of something as strange as light in everyday terms of solid, bullet-like particles, or as resem-

Lasers provide the most accurate method of measuring the Moon's distance. The narrow laser beam (blue) bounces off reflectors left by manned or unmanned spacecraft on the lunar surface, and the returning light is detected by the telescope (next to the laser) after its round trip of roughly 750,000 km (465,000 miles).

The entire electromagnetic spectrum, of which light is only a small part. The types of radiation differ fundamentally only in wavelength, shown by the lower scale (1nm= one millionth of a millimetre). The definitions overlap slightly; and the radio waves (right) are here shown divided into the conventional broadcasting bands.

In the insert, the dark lines show how the radiations from a hot body are distributed in the spectrum. A 6000°C body (like the Sun) emits mainly visible light, while a 2000°C object. emits mainly infra-red radiation.

bling ocean waves. Physicists can write down equations which describe how light behaves, and work out the consequences when it strikes a metal which can release electrons, or when it hits a diffraction grating. In the first case, the solution to the equations shows that light behaves as if it were a particle, and in the second as if it were a wave. There is no contradiction at the level of the basic equations, but it's impossible to comprehend in everyday terms.

The information in starlight
Looking at the fine detail in the spectrum of the Sun or other stars, we can see very narrow dark lines in the spread of colours. These are specific wavelengths at which the star's surface does not emit light, and in them lies the refutation of Comte's idea that we can know nothing of a star's make-up. Within a few years of his death, indeed, scientists had found that each chemical element produces a unique set of wavelengths, which acts as a spectral fingerprint for that element. And astronomers discovered that the dark wavelengths in star spectra corresponded to the wavelengths of light given off by elements such as hydrogen and sodium when studied in the laboratory. The element helium was first detected as a set of unidentified lines in the

Sun's spectrum, forty years before the gas was isolated on Earth.

When the structure of the atom was laid bare early this century, the distinctive fingerprint of each element could easily be explained. Each element has a different arrangement of electrons around it, which produces that element's unique chemical properties. But the electrons can only move in certain orbits – because of the now all-embracing quantum theory. If an electron moving in a large orbit, far out from the nucleus, wants to move 'down' into a lower-energy, smaller orbit, it has to 'jump'. Electron movements from orbit to orbit must go in jumps, each of which involves a certain exact amount of energy. If an electron jumps down towards the nucleus, the energy is emitted as a light photon of just that particular energy, and in the spectrum it appears as light of the corresponding wavelength. Similarly, an electron in this lower orbit can, when the atom is bathed in light of all wavelengths, remove light of this particular wavelength and use this quantity of energy to jump precisely into the larger 'upper' orbit.

A hot gas, whether in the laboratory, a street lamp, or as a hot gas cloud in space, consists of atoms continually colliding at high speeds. These collisions knock their elec-

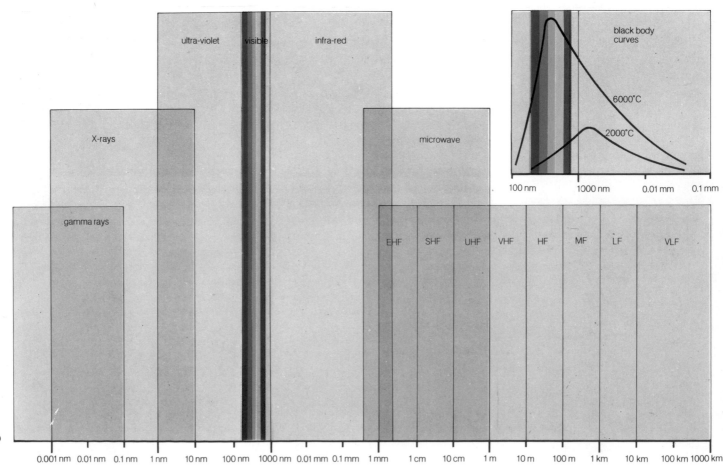

In holography, the beam from a
laser (right) is split in two. One
beam travels via the mirror
to the photographic
plate (near left); the other shines
on the object to be recorded
and the light reflected from
it also falls on the plate. Shining
a laser beam through the
developed plate (a hologram) then
produces a 3-D image of the
original.

trons into larger orbits; and when they jump down, they emit light waves at the 'fingerprint' wavelengths to give an *emission* spectrum, consisting only of bright lines on a dark background. A star's surface is different, however. Black-body radiation from its surface passes through the slightly higher and cooler layers of atmosphere, and the electrons of the atoms here pick out light photons which bump them up to larger orbits. So the star's original black-body spectrum of all colours is denuded of specific wavelengths, which appear as dark *absorption* lines in the bright continuous background spectrum.

Simply by identifying the wavelength patterns in the dark lines, astronomers can discover the various elements making up the outer layers of the star. Moreover, the types of lines also tell us the star's temperature. At high temperatures, atoms collide with more force, altering the distribution of electrons within them, and sometimes knocking electrons out of the atom altogether. In either case the emitted spectral lines alter, in strength or wavelength. Astronomers can measure temperature either from the spectral lines or from the star's overall colour: it's gratifying that the two methods agree.

The detailed way in which a star's dark lines intensify towards their central wavelengths tells astronomers of the star's surface gravitation, and of turbulence going on in its outer layers. But when measured precisely, the central wavelength of a line is never exactly at the wavelength that physicists find for the same line in the laboratory. For any star, all the lines are very slightly shifted toward either longer wavelengths (the reds) or shorter (the blues). The wavelength shift is easily understood when light is analyzed as a wave; and the effect was first investigated in the 1840s by the Austrian physicist Christian Doppler as a property of sound waves.

Doppler was intrigued by the fact that a sound seems higher in pitch when its source is travelling toward us, and lower as it moves away. In everyday life, we notice the sudden drop in pitch as the siren of a police car or ambulance passes us. Having worked out the theory, Doppler proceeded to test it. He arranged that a railway truck carrying trumpeters sounding various notes should be pulled at different speeds past a group of musicians with perfect pitch. The theory predicted a simple relationship between the pitch played, the pitch heard and the speed of the train; and his exotic experiment proved his theory to be correct. When the trumpeters are coming toward us, the sound waves, which cannot travel any faster than the normal speed of sound in air, are consequently 'bunched up', and we hear the sound as a shorter wavelength (higher pitch); while they are receding, since sound cannot travel any slower either, the note is stretched out to a longer wavelength (lower pitch). Similarly with light: if a star is moving

away from us the light waves are lengthened, and all spectral lines shift to the red; while an approaching star has a blue shift. The extension of a theory originally proved for railborne trumpeters now allows astronomers to measure a star's actual speed toward or away from us.

Lasers

There's one interesting side effect of the wave nature of light which has burst upon the world in recent years: the laser. A laser beam is powerful enough to cut through sheet steel and bricks; it is narrow enough to pick out reflectors left by astronauts on the Moon, and powerful enough for the reflection to be visible after the 750,000-kilometre (465,000-mile) round trip; and laser light has the strange property of 'coherence' which lets us take three-dimensional photographs. Yet despite its space-age connotations, the theory behind the laser was originated over sixty years ago – by that all-round genius, Albert Einstein.

Einstein was concerned at the time to tie together Max Planck's radiation formula with the then new theory of electron orbits in atoms. He proved that electrons must undergo three kinds of jump. Physicists already knew that the electrons could absorb light photons to jump up in energy, and that they could emit radiation by spontaneously jumping down. Einstein found another kind of downward leap, which happens when a photon with exactly the same energy as a possible downward jump passes the atom. The electron in the high energy orbit feels the 'stimulation' of

the passing photon, and immediately jumps down to give out a photon exactly similar to the stimulating one.

For over forty years this phenomenon was virtually ignored. But when a device for producing intense radio waves by a similar process (the *maser*) was perfected by American physicist Charles Townes, scientists raced to perfect the optical counterpart, an intense light source. By a narrow lead, another American scientist, T. H. Maiman, succeeded in constructing the first such device in 1960, and since then many versions have been made and marketed.

In principle, the theory is quite simple. First the electrons in the atoms making up the body of the device (an artificial-ruby cylinder in Maiman's original) must be moved up into large, high-energy orbits. We can put in enough energy by very bright lights around it, or suitable electric currents. Once most of the electrons are up in the large orbits, it won't be long – around one ten-millionth of a second – before one of them jumps down again quite spontaneously. And now Einstein's effect becomes crucial, for as the photon emitted by this electron leap moves out through the atoms, it will stimulate all the high energy electrons it passes to jump down and emit similar photons.

This build-up of light is easily enhanced by placing mirrors at each end, which force the photons to travel back and forth through the crystal, stimulating more and more electrons to emit. The legion of identical photons thus grows more and more intense, and a flood of light of just one exact wavelength builds up inside the device. We can release it either by an electro-optical 'shutter' at one end, or simply by making one of the end mirrors less than 100% reflective. The photon flood then emerges as a narrow, intensely bright light beam of just one wavelength.

The first of these devices (named 'lasers', for Light Amplification by the Stimulated Emission of Radiation) gave just a burst of light before the electrons had to be re-energized. Later models can produce a continuous beam of light, rather than a series of pulses. And another break-through on the doorstep is a laser which can be 'tuned' to any wavelength we choose, without being limited to those which occur naturally from electron leaps in particular atoms. Laser technology is a growth business, with no end yet in sight.

The intensity of a laser beam is largely due to the fact that all the light waves emitted by different atoms are 'in step' with one another. The stimulated emission process ensures that each new photon coincides with the original – here we have an excellent example of how physicists must describe light in terms of both particles and waves to make sense of Nature. As the laser process continues, all the photons end up in step – unlike an ordinary light bulb, which produces photons all independently, so that many of them just cancel out peak for trough with other photons.

Because laser light has this property of 'coherence', we can use it to take three-dimensional photographs. In this technique, *holography*, a beam of laser light is split in two, and one beam falls on to the object to be photographed. A photographic plate is positioned so that it is bathed in the second beam from the laser, and also picks up light reflected from the object (without any focusing). Since the object has shifted the peaks and troughs of the radiation which is reflected from it, in a highly complicated way, we find that the photographic plate has an intricate pattern of dark and light patches, where direct and reflected light have cancelled or reinforced each other.

Take the plate away, develop it, then shine a laser beam through it; and an image of the original object appears as a three-dimensional solid. You can look at it from different directions and see different aspects, as if it were real. The first holograms were made in just one colour, so everything appeared in monotone – often a ghastly green – but white-light holograms showing colours realistically are fast becoming a reality. Many museums are seriously considering replacing valuable exhibits with holograms, which from outside the display case would look identical to the original.

Lasers have important commercial and research prospects, too. By confining laser pulses to a very short time – a million-millionth of a second – enormously powerful light pulses can be produced, up to a thousand million million times as bright as sunlight. Such short, powerful pulses may eventually prove to be the key to fusing hydrogen nuclei into helium, and thus open the way to harnessing the hydrogen of the oceans to produce almost unlimited energy.

On a smaller scale, hospitals are now beginning to use lasers, particularly in eye surgery: a detached retina, for example, can be 'spot-welded' into place by laser pulses shone into the eyeball. And astronomers can measure the time taken for a laser pulse to return to the Earth after bouncing off the reflectors left on the Moon, to determine the shape of the Moon's orbit precisely. In this way they have already eliminated several rivals to Einstein's theory of gravitation (more of which in Chapter 9).

The longer waves

Light is far from being the only form of electromagnetic radiation. We have already hinted at the longer wavelength, infra-red; and the shorter wavelength, ultra-violet. The family of 'invisible' radiations is completed by the extremely long radio waves, and the ultra-short wavelengths called X-rays and gamma-rays. All these radiations are the same in essence. They all behave as both particles and waves of oscillating electric and magnetic fields, with the energy of a photon depending inversely on the wavelength.

The reason why we catalogue them under different names is really because we generate and detect them in different ways. Radiation can have any wavelength (and associated energy), but the units of matter, such as atoms, have definite sizes and energies; and so the various wavelengths interact with matter in different ways. A light wave finds itself a thousand times bigger than an atom, for example, but it has the right energy to force an electron from one orbit to another; while an X-ray's wavelength is atom-sized, but its higher energy means that it knocks electrons right out of their atoms. Because of these different interactions, the discovery stories of the various radiations differ widely. Invisible radiations beyond the ends of the Sun's spectrum were discovered accidentally with thermometers, in the case of infra-red, and photographic salts which darkened on exposure to the ultra-violet. German physicist Heinrich Hertz searched for radio waves after Maxwell's theory was published, while X-rays and gamma rays were accidental discoveries found during research on the conduction of gases and on radioactivity respectively.

The longest are radio waves. Any electromagnetic radiation with a wavelength longer than 1 millimetre (·039 inch) is a radio wave, as used for communications. The electric circuits in a radio transmitter force electrons to oscillate up and down a transmitting wire, or aerial, and the continuous acceleration and deceleration of these moving electric charges produces a continuous stream of electromagnetic waves from the aerial. The wavelength transmitted depends on the frequency at which electrons are forced up and down the aerial. To take another example, an ordinary electrical circuit is an alternating signal which forces electrons back and forth fifty times a second – a frequency of 50 Hz (the full name of the unit is 'hertz') in physics jargon. Any electrical appliance radiates electro-magnetic waves with the corresponding wavelength of 6000 kilometres (there's a simple relationship: wavelength × frequency = velocity of light), and this can be picked up by other wires nearby – as any hi-fi enthusiast with a hum in his amplifier knows! (50 Hz is the standard AC current in the UK, but in north America it is 60 Hz.)

Ordinary broadcasting uses much higher frequencies, and correspondingly shorter wavelengths: typically 900 kHz (900,000 Hz), 330 metres, in the medium wave band; or 90 MHz (90,000,000 Hz), 3·3 metres, in the VHF band. The longer wavelengths can travel long distances around the world – as we can easily find out by tuning through the medium wave at night – because waves travelling up-wards from the transmitter are reflected down again by a layer of electrons in the Earth's upper atmosphere. This ionosphere, some 300 kilometres (186 miles) up, acts as a mirror to long radio waves, and they can bounce back and forth between the ionosphere and the Earth's surface to cover surprising distances. (The improvement in reception at night is caused by atmospheric changes.) The very short

VHF waves, on the other hand, can only be picked up by receivers more or less on a line of sight from the transmitter, for they pass straight through the ionosphere into space. Local radio stations use VHF waves for this very reason, and any VHF programme is much freer of unwanted inter-ference from distant radio stations.

And because the ionosphere is transparent to radiation shorter than about 30 metres (100 ft), radio waves from natural sources in space can penetrate down to the ground. Radio astronomers study this radiation, just as 'optical' astronomers study the light coming from space, to help us understand what is going on in the depths of the Universe. Those celestial objects which are natural radio emitters are often, indeed, very faint in terms of light output, so radio astronomy has shown up – or drawn attention to – powerful objects which optical astronomers had missed: pulsars and quasars are the best known, but by no means the only examples in radio astronomy's half century of history.

The science arose by accident, when communications engineer Karl Jansky of the Bell Telephone Laboratories was investigating radio 'static' in the early 1930s. As well as the hiss from thunderstorms, he found radio 'noise' coming from the sky. Since then, radio astronomy has never looked back. Radio telescopes have grown larger, receivers more sensitive, and computers have been harnessed to analyze the results at a rate far faster than any human researcher. A radio telescope works just like any other radio set: it is basically an aerial connected to a receiver and amplifier. But sensitivity is the radio astronomer's nemesis. The total amount of energy picked up by all the radio telescopes in the world, over the entire history of radio astronomy, is less

Astronomical objects can emit the whole range of electromagnetic radiations, but only light and the shorter radio waves can penetrate Earth's atmosphere down to sea level. Infra-red astronomers work on high mountains, or from balloons or aircraft; for other wavelengths, an observatory in orbit above the atmosphere is required.

than you will exert in turning over this page.

The word 'radio telescope' conjures up a picture of a large bowl, supported and moved by a metal framework underneath. This type of 'big-dish' radio telescope concentrates radio waves by reflecting them on to the aerial above its centre, just as an optical telescope's mirror concentrates light. The receiver itself is especially sensitive – and costly. Some are cooled in liquid air for greater sensitivity, while others incorporate a maser (the microwave equivalent of the laser) to amplify the tiny radio signal from the sky. And receivers are designed with the peculiarities of cosmic radio source in mind. Some sources emit at just one wavelength – like a man-made transmitter – and these correspond exactly to spectral lines in visible light. Astronomers study these 'radio lines' in detail to work out how the invisible gas in our Galaxy, and others, is moving. Most of the radio sources, however, produce all wavelengths of radio waves, and it's not too important where the radio astronomer 'tunes in'.

The radio astronomer also has problems in seeing fine details in a radio source. Any telescope is limited in theory in the details it can see, and a large telescope can naturally see finer details than a smaller one. Optical astronomers are not too concerned, for their limit is a much coarser one, set by the fluctuations in the atmosphere. But radio astronomers find the natural limit, set by the wave nature of radiation, a real handicap. Their wavelengths are roughly a million times longer than light, and telescopes have to be that much bigger to see the same amount of detail.

Britain's Astronomer Royal, Sir Martin Ryle, has pioneered an ingenious solution: connect together electronically several small radio dishes set out along a straight line, and let the Earth's rotation swing them around. With the aid of a computer to analyze the results, it's possible to achieve the resolution of a telescope as wide as the aerial array is long. His 'Five Kilometre' telescope at Cambridge, built along the line of the former Oxford to Cambridge railway, can 'see' the same amount of detail as an optical telescope, despite the much longer wavelength of radio waves.

An even stranger approach is to record signals from individual radio telescopes on opposite sides of the world, both looking at the same source at the same time. By combining the results later, radio astronomers have glimpsed some of the detail which would be provided by a radio telescope as wide as the Earth – detail equivalent to the width of a human hair at 10 kilometres (six miles) distance!

Light's near relatives
Shorter than radio waves are the infra-red wavelengths,

115

familiarly known as 'heat' radiation. They fill the gap between radio waves and the red end of the spectrum of light, and it's here that the radiation from warm and moderately hot objects peaks. An electric heater gives out a vast amount of infra-red, and not many light or radio waves; and this radiation vibrates the molecules in our skin, so making us feel warm. Under a grill, or broiler, the radiation is intense enough to break up molecules, and so cook our food.

This is fairly wasteful of energy, though, since the heat has to penetrate inwards, taking time and possibly burning the surface. Microwave ovens take an entirely different approach, and cook the food from within. These waves are on the borderline between radio and infra-red, and a very powerful microwave beam can be generated electronically. The wavelength is chosen to match the natural rate at which water molecules vibrate, and since food is largely made up of water, these molecules absorb the microwaves passing through them, and heat the food right through. The result is quick and thorough cooking – and the plate, which contains no water molecules, is left cold.

Infra-red astronomers pick up radiation from relatively cool stars and from warmish clouds of dust in space. But our atmosphere blocks off most infra-red wavelengths, which are absorbed by water and carbon dioxide molecules in the air. Astronomers studying the infra-red must work at high-altitude observatories, or from balloons and high-flying air-craft. And their task is one of the hardest of all astronomers', for the surrounding air and the telescope itself are at just the right temperature to be emitting their peak radiation at the wavelengths at which the astronomer is looking. It's as if an optical astronomer had to contend with a telescope painted inside and out with phosphorescent paint! Advanced electronics can get around this problem to some extent.

Modern optical astronomy, too, is becoming increasingly an electronic wizard's paradise. Although photography is still extremely important, electronic methods of detecting the individual photons of light make it easier to measure brightness exactly, and to get really accurate spectra. Computers are regularly used to guide telescopes, to analyze results, and to assist in measuring positions of photographic plates. Astronomers' enormously increased understanding

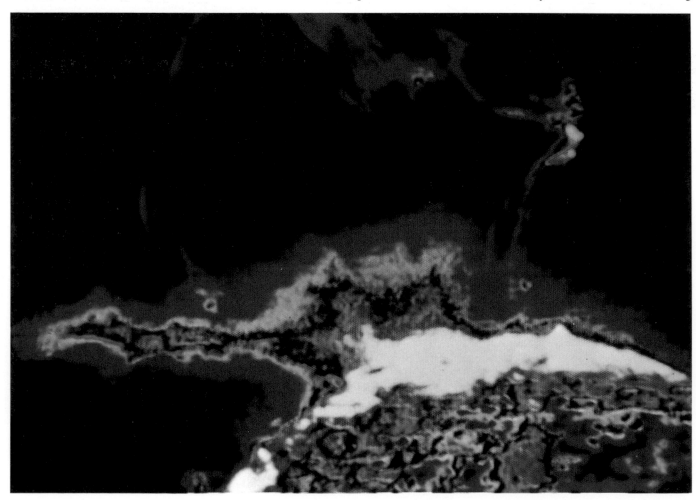

of the Universe in the past couple of decades has been as much due to the computer as to any other single factor.

At wavelengths shorter than that of light, the Earth's atmosphere again blocks off all signals from space. Astronomers talk in terms of two 'windows' in the atmosphere, one of which lets in light, the other radio waves. At all other wavelengths the sky is an opaque veil.

Down beyond the short (violet) end of the spectrum lie the ultra-violet rays. Individual photons of these radiations are energetic enough to break up molecules, and the unpleasant condition of sunburn is just a reminder that without the ultra-violet shield of our atmosphere, life just could not exist on land. The Sun's ultra-violet rays are blocked by a layer of ozone – oxygen molecules containing three instead of two atoms – some 50 kilometres (30 miles) up. Recently, environmentalists have pointed out that the gas from aerosol cans, or from high-altitude aircraft exhausts, may break up the ozone molecules. If this is a serious possibility, it threatens all life on the surface of Earth with a deadly flood of solar ultra-violet rays.

Ultra-violet astronomy must be conducted from remote-controlled satellites orbiting well above the ozone layer. They pick up very hot stars, whose radiation peaks in this part of the spectrum; and a detailed analysis of the star's ultra-violet spectrum reveals absorption by molecules in intervening space. The most important of these is the hydrogen molecule – the normal Earthly form of two joined hydrogen atoms – for although hydrogen atoms in space had been discovered by radio astronomers, hydrogen molecules were unexpected; astronomers now think that half the gas in our Galaxy is in this form.

Exotic radiations

Moving down to the still shorter X-rays, whose wavelengths are roughly equal to the size of an atom, we find the photons still more energetic, more disrupting and more penetrating. In medicine, penetrating X-rays are invaluable in showing up internal disorders in the human body, and the latest technique of *tomography*, scanning with an X-ray source, gives a three-dimensional view of brain tumours. But there's always a very slight risk of damaging the body tissue (large X-ray doses are actually used to destroy cancerous tumours), and doctors are now beginning to turn to completely safe ultrasound and magnetic body-probing techniques.

X-rays from space generally come from extremely hot gas clouds, at temperatures of thousands of millions of degrees. Satellite observatories have located the positions of some three hundred of these X-ray sources, and these high-temperature objects (usually invisible to ordinary optical telescopes) are telling us of some of the strangest objects in the Universe – such as the black holes, which we will meet in a later chapter.

Finally, the shortest radiation of all is the gamma-rays. There's no hard and fast dividing line between ultraviolet and X-rays, or between X-rays and gamma-rays. On Earth, the distinction usually depends on how we produce them, but when we are picking up signals from space the dividing wavelengths are quite arbitrary.

Gamma-ray astronomy is still a very young science, and little is known of celestial gamma-ray sources. The first gamma-ray satellites were the American Vela series, actually a classified military programme for detecting nuclear explosions on Earth. They began to pick up sudden bursts of gamma-rays in 1967, and it was soon obvious that they were coming from space rather than from our planet. The first explanations for them ranged from comets crashing onto relatively nearby ultra-compact neutron stars, to exploding stars in distant galaxies; but it now seems likely that they are related to ordinary celestial X-ray sources, which we'll be looking at in Chapter 9.

From radio waves to gamma-rays – spanning a range of a million-million-million-fold difference in wavelength – today's astronomers reap the electromagnetic radiation from the Universe. The complex radio sets of the radio astronomer, the giant Earth-based telescopes of the optical astronomer, and the orbiting satellites which astronomers must use to study other wavelengths, all are filling in our picture of what's going on out there.

We can pick up other signals from space which are not electromagnetic, but they are very few. Fast-moving charged particles – electrons, protons and various atomic nuclei – smash into the top of our atmosphere as 'cosmic rays' after a long journey from exploding stars. The elusive neutrinos are another kind of particle from space, but they can only be trapped with difficulty. Astronomers have so far detected only neutrinos from our nearest star, the Sun.

More akin to electromagnetic waves are gravitational waves, tiny fluctuations in the gravitation of distant objects as they spin wildly or collapse catastrophically. American physicist Joseph Weber pioneered the search for gravitational waves, using two large suspended aluminium cylinders each weighing over a ton. He looked for sudden vibrations of the cylinders, caused by an outside 'hammer blow' of some kind. Earthquakes could obviously do this, so he installed the cylinders at well-separated sites near Washington and Chicago. The only 'blow' which could affect both cylinders simultaneously is a burst of gravitational waves from space, and his first results seemed to show many of these sudden simultaneous vibrations. Unfortunately, other researchers have since failed to find this plethora of gravitational bursts, but Weber's lead has encouraged others to build more sensitive equipment, and possibly gravitational waves will someday be discovered.

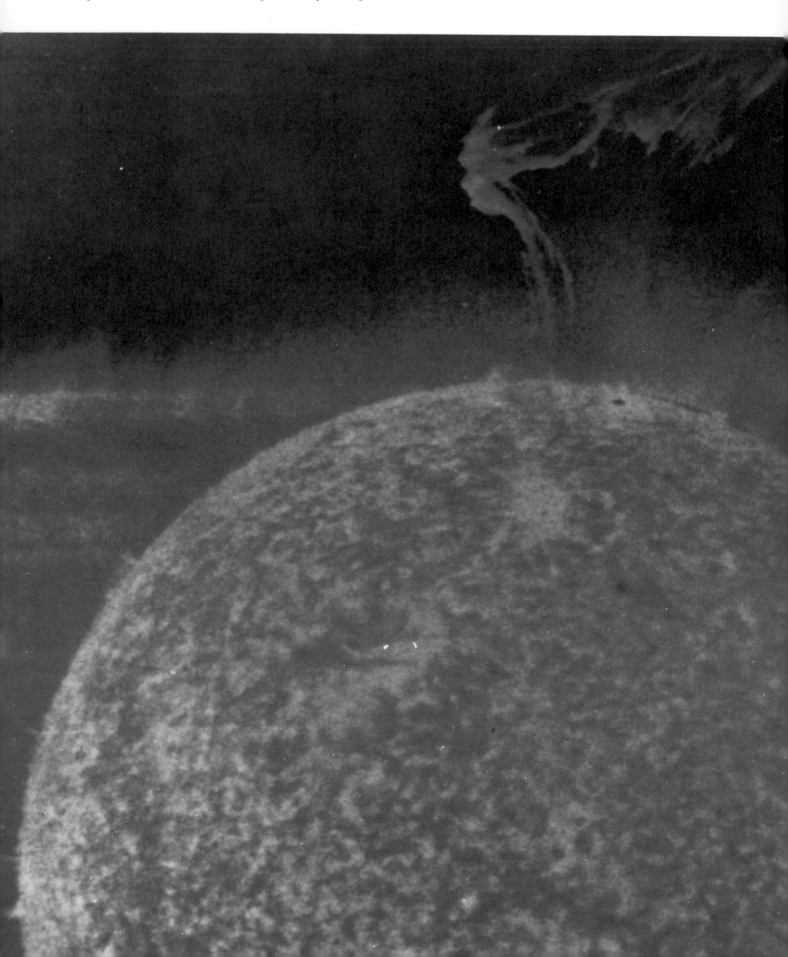

A huge eruption on the Sun, our local star, photographed in X-rays by the Skylab astronauts. The false colours show temperature variations, while a row of overlapping fainter images causes the bright streaks and patches to left and right.

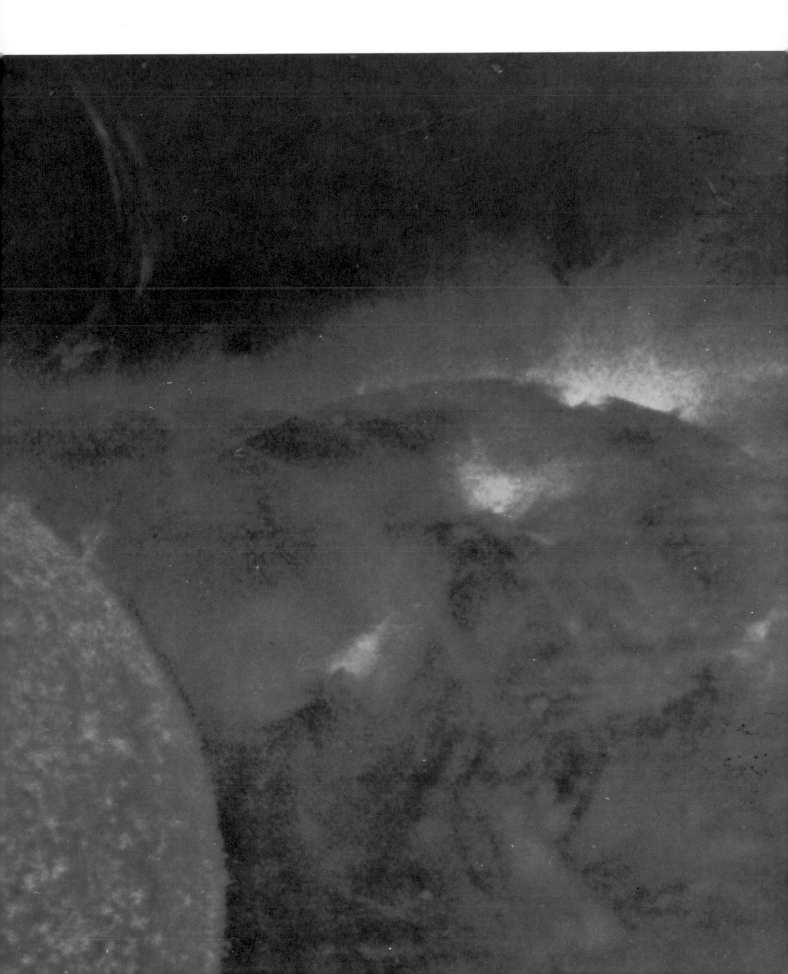

Two thousand metres (6,500 feet) below ground level, near the South Dakota town of Lead, an astronomer is seeing into the heart of the Sun – with a 400,000-litre (88,000 gallons) tank of cleaning fluid. Raymond Davis is trying to catch a few of the ghostly neutrinos which pour out from the Sun's infernally hot core as hydrogen nuclei fuse into helium.

The flux (rate of energy transfer) of solar neutrinos is immense. They carry one-fiftieth of the energy leaving the Sun, and they are streaming through your body at the rate of a hundred million million every second. Yet they are remarkably elusive, because they are virtually unstoppable. With no electric charge, they are unaffected by the electrons and nuclei of the atoms they pass through; and their velocity, virtually equal to that of light, ensures they are deflected only slightly by gravity. They are immune to the strong nuclear force, and only give away their presence by undergoing the aptly-named weak interaction – a typical neutrino could travel through light years of lead shielding before being stopped.

Davis' tank is filled with tetrochloroethylene, a liquid rich in chlorine atoms – and one in four of the latter is the

This page, below :
The nineteenth century German scientist Hermann von Helmholtz studied topics as diverse as the human eye, sound waves and

energy. He calculated that the Sun's gravitational energy could keep it shining for only 30 million years.

This page, top :
Computers assist modern astronomers in calculating how stars change as they grow old. Results are conveniently displayed

as a graph : this example reveals (to the trained eye) how a double star system alters as one star expands in its old age.

heavy variety, chlorine-37. Occasionally, a passing neutrino is stopped by a chlorine-37 nucleus, the impact shattering one of neutrons in the nucleus into a proton and an escaping electron. The chlorine-37 is thus transmuted into an atom of the inert gas argon-37. Every few months, the huge tank is flushed out with helium gas, and the flushing gas is searched for signs of the tell-tale argon. The unusual choice of location, deep in the Homestake gold mine, arises from the need to protect the chlorine nuclei from other transmuting radiation and particles. The 2000-metre umbrella of rock shelters the tank from the continuous rain of cosmic rays – very fast nuclei and electrons from the depths of space – while being virtually transparent to the insubstantial neutrinos.

This seemingly esoteric experiment, which has been running for ten years, is a vital check on the theories of the life cycles of stars. Over the last forty years, astrophysicists have refined their theories of stellar evolution to a remarkable degree: it is possible to work out fairly closely the temperature and mix of elements in the centres of various types of star, to calculate how long a star stays at each stage of its evolution, and to predict how a star will eventually die. But there are very few direct checks on these calculated results. The entire gamut of electromagnetic radiations we receive from a star, from X-rays through light to radio waves, comes from its surface layers and atmosphere. Neutrinos, on the other hand, shoot straight out into space from the central nuclear reactions themselves. The theorists' models of stars predict just how many neutrinos should be produced; Davis set out to check the calculations as they apply to our local star, the Sun.

The star builders

At first it may seem mysterious that astronomers can know anything about the interiors of the remote stars; after all, they appear as mere points of light in even the largest telescopes (unless special techniques are used). As we saw in Chapter 7, however, the spectroscope reveals the elemental make-up of a star's outermost layers, and this spectrum – or just the star's colour – tells of its surface temperature. The total luminosity can be calculated if a star's distance is known, and the mass if it is one of a gravitationally bound pair (see Chapter 2).

Using a large computer, a modern astrophysicist can build a huge range of theoretical 'model' stars. He uses the example of a lump of gas of some particular mass and a specified mix of elements, lets it settle together through gravitational attraction, and follows changes in temperature, pressure and composition throughout the star as nuclear reactions and other processes take place. A formidable task indeed; but in principle it is just a matter of applying the known laws of physics, no different from calculating the

path of a ball thrown into the air.

The Universe, however, must be the ultimate touchstone. A theorist's model star may end up not corresponding to the intended real star, either because our laboratory physical laws do not hold in the extreme inferno of a star's

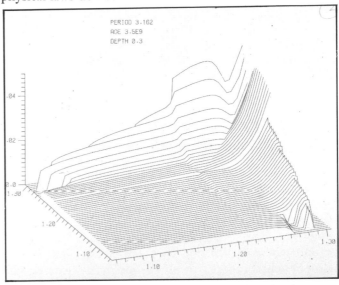

heart, or simply because some of the calculations have been too simplified, allowing for easier and quicker computing, but perhaps giving an inaccurate answer. The first of these possibilities led to a major crisis when physicists first made calculations on the Sun's energy output in the mid-nineteenth century, before the power locked up in atomic nuclei was known. The great German scientist Hermann von Helmholtz realized that gravitation was the only force then known which could keep the Sun hot for a long time. A gas cloud shrinking as it pulls itself together by gravitation heats up, but a simple calculation showed that the energy produced would only allow the Sun to shine at its present rate for 30 million years. Geologists declared this result to be impossible. The Earth and, more important, fossil remains of life on Earth, were far older than this. Not until the 1930s was the problem finally resolved, when another German, Hans Bethe, deduced the nuclear reactions which can occur in the searing heat of the Sun's core – reactions completely outside the laws of physics known in Helmholtz' time. The advance of laboratory physics had allowed astronomers to explain the stars' luminosity, and modern astrophysicists can pay back the debt: by treating the Universe as a huge laboratory, where Man can see the results of Nature's own experiments on a scale impossible to achieve on Earth.

The results of Nature's 'experiments' in star building can be conveniently summed up by the mass-luminosity and the temperature-luminosity (Hertzsprung-Russell) relations, which we have already met in Chapter 2. These two are the Rosetta Stone of the human 'star builders'. Theoretical 'model' stars must obey the relations, and it should also become obvious why most stars lie on the 'main sequence', while a few are red giants or white dwarfs. In fact the calculations agree with the relations found for real stars, and astronomers have every confidence that if they could actually dig into a star, its interior conditions would turn out to be in agreement with the theorists' models.

Perhaps more surprising, the changes a star undergoes during its life are well understood, too. Stars' lifetimes range from only a million years to over 100,000 million years, and most changes are correspondingly slow. Occasionally astronomers have seen a star 'switch on' and begin to shine, while a sudden star death may be signalled by a distant supernova explosion, but generally the stars seem unchanging. Our situation is like that of an extraterrestrial spending an hour in one of Earth's busy streets: although he does not see any passer-by age appreciably, he can compare individual humans in the crowd, deduce what characteristics signify age, and hence find how humans mature from infancy to old age. Similarly, the theories of stellar structure indicate what changes we should expect in a star as its central nuclear reactions convert one element to another, and so we can divine the ages and the life stories of the crowd of stars around us – including the past and future of our own Sun.

Before describing the main, and best understood, stages of a star's life, however, it is time to return to the birth of stars, a process which until recently has literally been shrouded in obscurity.

Star birth

Star 'nurseries' are hidden from astronomers' eyes, in the depths of huge clouds of gas and obscuring dust. Gas clouds like the famous Orion Nebula fill much of the plane of our rotating Galaxy. An interstellar cloud is generally quite stable, as gas pressure trying to spread the cloud into the thinner gas around it is counterbalanced by the force of its own gravitation holding it in. But the balance is tipped if the cloud is suddenly compressed. Perhaps it crashes into

just as Helmholtz calculated for the 'shrinking Sun' last century; and as the 'surface' – in reality, a cocoon of thicker dust around the star – reaches a temperature of a few hundred degrees, they begin to glow dimly. Like an electric heating element at a similar temperature, most of the energy comes off as infra-red (heat) radiation. Astronomers have found dozens of infra-red cocoon stars clustered in the centre of clouds like the Orion Nebula.

In the late 1960s Eric Becklin and Gerry Neugebauer discovered a particularly bright infra-red star in this nebula, behind the glowing gas and stars which show up on ordinary photographs. The Becklin-Neugebauer object is probably a proto-star some ten times heavier than the Sun, and glowing ten thousand times brighter – if it were not cocooned and obscured by dust, we would see it as one of the brightest stars in the sky. Instead, the Becklin-Neugebauer object is invisible to even the largest optical telescopes; and the infra-red properties suggest that only one part in a million million million of its visible light can escape through the cocoon and surrounding nebula.

Radio astronomers have no less exciting news. For a long time they have mapped the hot hydrogen gas near young stars, but detailed examination of the *spectrum* of radio waves from dense gas clouds has recently shown the tell-tale 'lines' of many molecules. Previously, only single atoms (and the electrically charged atoms called ions) had been found in space, but nearly fifty different chemical compounds have now been detected in interstellar clouds, and the list is growing every year. These molecules assemble from atoms of gas on the surface of the tiny dust grains, and the dust protects them from the ultra-violet radiation which normally disrupts molecules in space. Some of the larger molecules are intriguingly similar to those found in living cells, and some scientists suggest that these interstellar chemical reactions may play an important part in triggering off life on planets which eventually form in orbit about the proto-star.

Even more unexpected, some of these molecules are in small clouds which shine out brilliantly at radio wavelengths, as natural *masers* (the radio equivalent of lasers). Each is about the size of the orbit of the Earth around the Sun, and is powered by the infra-red radiation of nearby proto-stars, converted by molecules of water (H_2O) and hydroxyl ions (OH) into intense microwave beams. They assist star birth by acting as cosmic refrigerators, drawing out heat which would otherwise oppose the gravitational shrinking of the proto-stars.

The stars' power house
The central temperature of the proto-star soars continuously during its collapse; when it reaches the colossal temperature of ten million degrees, nuclear reactions begin.

one of the spiral-arm gas clouds as it drifts around the Galaxy's centre, or it may be hit by the expanding debris from a nearby supernova explosion. Either way, the cloud is flattened into a cosmic pancake, and gravitation takes hold to pull the gas into shrinking clumps. As the clumps contract, they spin faster, obeying the law of angular momentum conservation, which we met in Chapter 3 (just as a spinning ice-skater spins faster as she draws in her arms), until eventually parts break away to form independent condensations. Hence a whole cluster of shrinking 'proto-stars' forms, hidden inside the remains of the original cloud.

But in the last few years, astronomers have acquired eyes to peer through the 'nursery curtains'. Infra-red rays and radio waves can penetrate the smoky dust clouds, and astronomers can now pick up these radiations and 'see' the proto-stars and their surroundings. The shrinking proto-stars warm up as their gravitational energy turns to heat,

The hydrogen nuclei (protons) fuse together in fours to make up helium nuclei. To start the reaction, the nuclei must hit with sufficient force to overcome the repulsion due to their positive electric charges – 'like charges repel' – and, since temperature is a measure of the speed of the gas particles, a high enough temperature will ensure that fusion begins. (This high temperature is the most serious problem in designing a fusion reactor on Earth, by the way.)

As the strong nuclear force binds the particles together, energy is carried away by the high speed of the products, and by gamma-rays. The reason for the energy loss can be seen quite easily by considering the reverse reaction: a large amount of energy would be needed to decompose a helium nucleus to two free protons and two free neutrons, and an exactly equal amount of energy is therefore released when the components of a helium nucleus bind together. It is this loss of binding energy as nuclei combine which powers the stars. All forms of energy have a certain amount of mass associated with them, according to Einstein's famous equation $E = mc^2$ (energy = mass × speed of light squared), and the loss of energy means that each helium nucleus is only 99·3% as heavy as the four hydrogen nuclei from which it was formed. A typical star like our Sun converts mass into radiation at the staggering rate of 4 million tonnes every second!

The star is literally centrally heated by the energy lost from the fusing nuclei in its core (apart from the small amount of energy carried away straight into space by the elusive neutrinos). The main reaction products – fast nuclei and particles, and gamma rays – cannot travel far without meeting particles of the surrounding gas; and here the electromagnetic force between them comes into play, slowing down the reaction debris and cannoning the gas particles up to high speeds. The average speed of the particles thus increases, constituting a further temperature rise in the star's already superheated core. But heating a gas increases the pressure it exerts on its surroundings – as anyone who has opened a warm champagne bottle knows – and this extra pressure at the centre of the star opposes the ever-present inward pull of gravitation. The star stops shrinking, and the perfect balance of forces keeps it shining steadily for millions of years.

And what of the fourth force, the weak nuclear force? It, too, plays a vital role. As the four hydrogen nuclei combine to make up a helium nucleus, two of them must change from protons to neutrons. This they accomplish by ejecting a positron (antielectron) and a neutrino – a weak nuclear reaction. As we have seen in Chapter 3, the weak reactions are about 100 million million times *slower* than the strong nuclear reactions; and the necessity for weak processes in the chain of helium-building reactions moderates the rapid assembly of helium nuclei. In the so-called 'hydrogen'

bomb, the explosive is not ordinary hydrogen, but heavier varieties whose nuclei contain one or two neutrons in addition to the one proton; and, as there is no need to wait for the weak process to generate neutrons, the strong force carries all before it: the nuclei fuse together in an almost instantaneous cataclysm. If protons could turn to neutrons at a rate corresponding to the strong interaction, the Sun would have blown itself apart as soon as it formed, as a cosmic-scale hydrogen bomb.

A steadily shining star like our Sun is thus a titanic battleground between the natural forces. On the one hand, gravitation is continuously trying to shrink it to the smallest possible size, and left to itself would succeed in roughly Helmholtz' time scale of 30 million years. But the strong reaction is binding together nuclei at the star's core, and releasing energy at a rate moderated by the weak nuclear force. Electromagnetic forces ensure that this energy heats up the interior of the star to produce an outward pressure which exactly balances gravitation.

Main-sequence stars

Not surprisingly, this condition of balance ensures that all hydrogen-fusion stars of the same mass are like identical twins, with remarkably similar properties. Suppose a star

Below:
The Eta Carina Nebula, a glowing
cloud of gas crossed by dark lanes
of dust. This photograph, from the

Cerro Tololo Inter-American
Observatory, Chile, is peppered
with hundreds of stars, each a
nuclear powerhouse like our Sun.

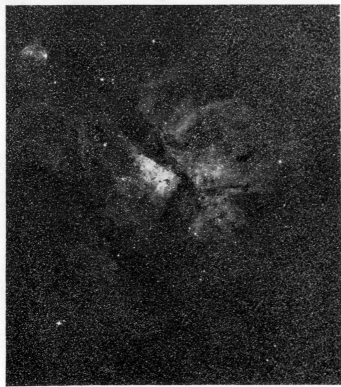

were producing less energy than is appropriate for its mass, for example. The centre would become too cool, the decreased gas pressure there would then allow gravitational forces to start to shrink the star; this contraction would heat up the core, and the higher temperature would increase the rate of nuclear fusion, with a corresponding return of energy production to normal. This superbly tuned feedback system ensures that stars shine more or less constantly over the time that they are fusing hydrogen to helium.

The Sun is roughly halfway through its 10,000-million-year hydrogen-fusion period; but, paradoxically enough, less massive stars will outlive it. Although these stars have lower fuel reserves, their smaller gravitation gives them a lower central temperature, and the fusion rate is so much slower that their hydrogen supply will in fact outlast the Sun's. In a collapsing gas cloud there will be proto-stars of all masses, and calculations show that only those heavier than one-twentieth of the Sun's mass (fifty times Jupiter's mass) will reach a high enough central temperature for hydrogen nuclei to fuse together. A less massive ball of gas will warm up by contraction, but it will never shine in its own right, and it will end up as a large planet – like a more massive Jupiter. True to this theoretical prediction, the lightest star known – Ross 614B – is some seventy times heavier than Jupiter, just above the limit.

The more massive stars are correspondingly lavish in their use of hydrogen. A star ten times heavier than the Sun lives for only a thousandth as long (ten million years) as nuclear reactions in its superhot core rip through the hydrogen fuel. More massive stars are thus expected to be more luminous than the Sun, and the calculated relation is very similar to the mass-luminosity relation for real stars. The high luminosity makes the most massive stars unstable, the huge outward flux of energy-carrying photons from the centre blowing off their outermost layers, even against the enormous pull of gravitation. The heaviest stars known, like the overweight pair called Plaskett's star, are around sixty times heavier than the Sun, and close to this upper limit.

The theory also relates the luminosity of a star to its surface temperature, and the prediction corresponds very well with the main sequence of the Hertzsprung-Russell diagram. Main-sequence stars are thus hydrogen-fusion stars, and the position that a star occupies on the sequence depends basically on its mass. As we shall find later, the other rarer types – including red giants and white dwarfs – are simply older stars which have already gone through the main-sequence phase. Most of a star's life is spent fusing hydrogen, and so most stars we see in the sky are of the main-sequence type; just as the great majority of the people we see in the streets are under the age of 60, simply because a human spends more of his life below the age of 60 than above it.

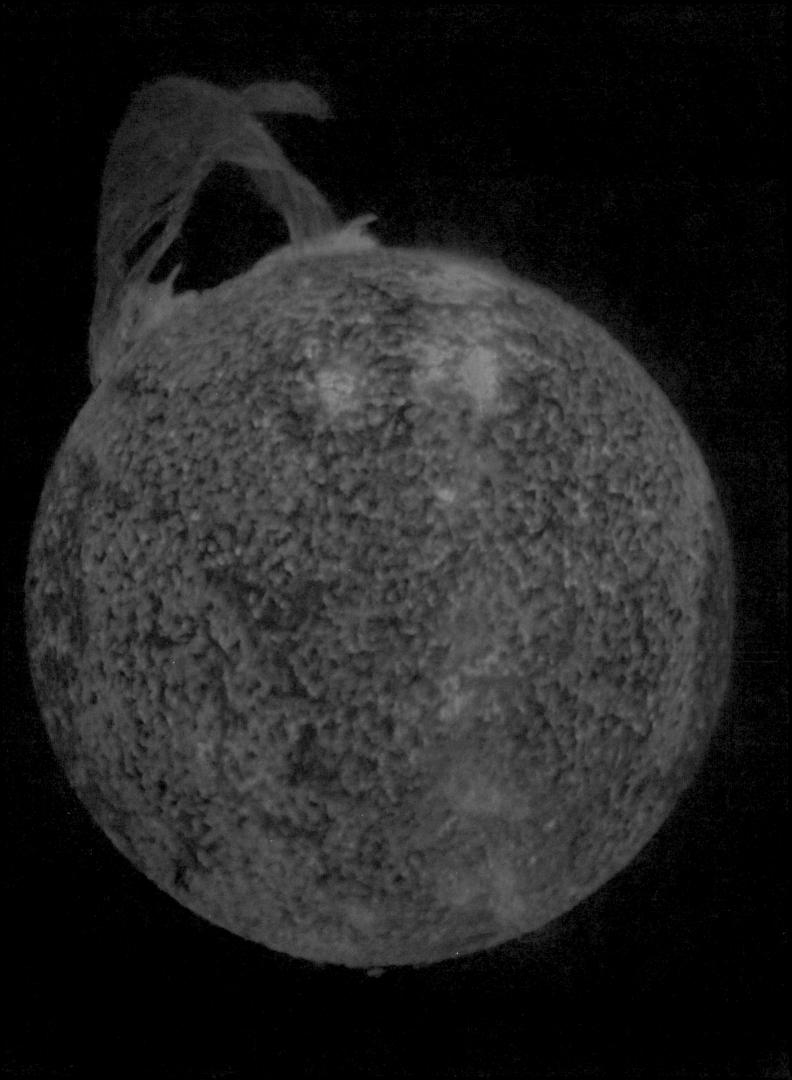

The close agreement between calculation and the observed mass-luminosity and temperature-luminosity (Hertzsprung-Russell) relations shows that the theory of hydrogen fusion is pretty accurate. Before following through the later stages of a star's life, after it reaches the crisis point of having converted all its central hydrogen to helium, let us turn to the only hydrogen-fusion star we can study in detail.

The Sun

The bland, smiling face of the Sun depicted by artists since ancient times is a good representation of how it hides most of its secrets from the enquiring astronomer. Until Raymond Davis began searching for neutrinos in 1968, there was no direct way of exploring its hidden interior, and our knowledge of its structure is largely based on the theory which so well describes hydrogen-fusion stars as a whole.

The energy from the Sun's core percolates outwards very slowly. Each individual photon travels only a fraction of a millimetre before it meets a gas particle and is absorbed or scattered. Each time, its direction and energy are changed. We saw in Chapter 7 how the character of electromagnetic radiation (photons) depends on the temperature of the surrounding matter. As the energy diffuses out from the extremely hot core, the principal type of photon changes from X-ray to ultra-violet to visible light by the time the energy emerges from the Sun's shining surface. This slow diffusion rate means that energy produced now will take millions of years to traverse the 700,000 kilometres (434,000 miles) to the Sun's surface – although this distance in empty space would be covered in just over two seconds!

Just below the Sun's visible surface is a zone where the outward energy flow is carried by the up-and-down flow of huge gas columns. A heated saucepan provides an excellent illustration of this convection process. Before the water in the pan boils, there is a stage when hot water columns are rising to the surface, while adjacent denser cool water sinks, giving a honeycombed surface of convection cells. The Sun's surface is covered with such cells – resembling a bowl of rice, each 'grain' showing a hot, bright centre of rising gas, and cool, dark edges where gas sinks back into the Sun after giving off some of its heat. These 700-kilometre (400-mile) diameter granules are arranged into larger (30,000-kilometre, 18,600-mile) super-granulation cells, seen at their best in the Sun's lower atmosphere by means of specialized instruments. We shall see below that these supergranules are intimately involved in the Sun's weather.

It may seem contradictory to talk about the Sun's 'surface' and 'atmosphere', for the Sun is gas throughout, becoming less dense from the centre outwards. The outermost parts of the Sun's tenuous atmosphere – outward-streaming gas called the solar wind – reach out beyond the

orbit of Jupiter: in a sense, most of the planets are orbiting within the Sun. These outer layers are extremely tenuous, though, and more rarefied than the best 'vacuum' we can produce on Earth. The idea of a definite surface is in fact quite reasonable, for photographs show the Sun to have a very sharp edge when seen in ordinary light. (But never look at the Sun directly through a telescope or binoculars: the concentrated heat can damage your eyes permanently.) In the region of the Sun's surface (photosphere) the gas density drops a thousand times over only a few hundred kilometres – a remarkably thin layer when seen from the distance of the Earth. Beneath the photosphere, the Sun's gas is dense enough to be opaque, while above – in the 'atmosphere' – it is transparent to light.

Magnetic weather

At first sight, the Sun seems to be the simple hot ball of gas that theory predicts. But that is because the huge flux of energy from the core emerges in the visible-light region of the spectrum, and in ordinary photographs it completely drowns out the phenomena of the Sun's atmosphere. If we look at the Sun in terms of radio waves, or X-rays, or in the narrow 'lines' in the visible spectrum which come from particular atoms and ions, the Sun's weather is revealed as a veritable maelstrom of activity. Scattered over the surface are huge *active regions*, some 150,000 kilometres (93,000 miles) across, marked by hot clouds in the lower atmosphere which shine brightly in light emitted by hydrogen and calcium gases. Feathery prominences stand out from the surface, while the sudden explosion of a flare occasionally shoots out a spray of gas into space. Above all this float huge

The glowing gas in this prominence, visible at the Sun's edge, shows up loops in the solar magnetic field, much as a bar magnet's field can be traced by scattering iron filings around it.

Below:
The Sun's outer atmosphere, the corona, can only be seen when the Moon hides the Sun's bright disc

during a total solar eclipse. This eclipse was photographed from Miahuatlan, Mexico, on 7 March 1970.

concentrations of gas in the upper atmosphere, showing up as brilliant patches in photographs taken of X-ray emissions.

All these 'weather patterns' of the active region arise entirely from the Sun's magnetic-field activity. The general field is probably generated by internal electric currents, rather like the Earth's magnetic field; and, despite the Sun's huge size, it is no stronger than the Earth's. But the gaseous Sun does not spin in a simple way, like the Earth's rocky ball. Its equatorial regions are rotating faster than the rest; and they 'wind up' the magnetic field, wrapping it around inside the Sun like an elastic thread. The bubbling convection layer beneath the Sun's surface twists the field into 'ropes' some hundreds of times stronger than the general north-south field. These ropes come up at the edges of the big supergranules, and show up in the Sun's atmosphere as grass-like spikes when photographed in the light from hydrogen.

Larger magnetic concentrations produce the active regions, trapping hot clouds of gas in the lower atmosphere, and concentrating the tenuous upper atmosphere into the X-ray-emitting regions above. This even hotter gas can gradually cool, and flow down the magnetic field to show up as the long, glowing arcade of a prominence. And where oppositely directed magnetic fields meet, they annihilate

each other with the furious explosion of a flare.

In the centre of an active region, the strongest part of the magnetic field causes the only phenomenon which shows up on ordinary photographs of the Sun: sunspots. The largest

Aurorae are more frequent when the Sun is at an active part of its eleven-year cycle. Historical records of bright aurorae allow astronomers to follow past solar activity.

are visible to the unaided eye when the Sun rises or sets through mist (again, never look at the Sun directly) and they were known to the Chinese as early as 43 BC. Galileo began the scientific study of them in 1610, and his insistence that the supposedly perfect Sun was marred by dark spots was one of the points which brought him into conflict with the Church authorities.

A sunspot is only dark by contrast, because it is cooler than the rest of the photosphere – and 'cooler' still means a temperature of some 4,000°C (7,000°F). If we could block off the blinding light of the nearly 6,000°C (10,000°F) photosphere, the spots would shine out orange red. Modern theories show that the intense magnetic field in the spots – several thousand times the Earth's field – will naturally cool the surface where it passes through.

Centuries of sunspot observations have revealed that the number visible on the Sun comes and goes in a fairly regular eleven-year cycle. Every eleven years, when sunspot numbers reach a minimum, the Sun's general field switches over, the N magnetic pole becoming an S for the following eleven years. After each reversal, the new general field is gradually 'wound up', and as a consequence new active regions (with their associated sunspots) begin to appear. After four or five years, a maximum of activity is reached,

for the strong magnetic fields of decaying active regions are spreading out over the Sun's surface and beginning to neutralize the general field. After eleven years are up, they have succeeded to the extent of reversing the general field, and another cycle begins.

The Sun's 'weather' affects ours

Since our Earth is orbiting within the Sun's outermost atmosphere, it is not surprising that we can detect signs of this eleven-year cycle on Earth. The radio-reflecting layers in the Earth's ionosphere are affected by the charged particles streaming out from the Sun's active regions; and radio reception changes markedly with the solar cycle. The beautiful, shimmering curtains of the Northern and Southern Lights – the aurorae – are atoms of our atmosphere stripped of their electrons by solar particles channelled in by the Earth's magnetic field. The number and brightness of the aurorae follow the solar cycle too; and mediaeval records of auroral displays are used to plot solar activity backwards for centuries before sunspot observations were regularly made.

But the eleven-year cycle also shows up in more surprising circumstances. The quality of wine vintages, and the number of 'centuries' scored in first-class cricket both

During the late seventeenth century Europe experienced winters so cold that the Thames froze over in London, and the citizens held

fairs on the ice. This 'Little Ice Age' was probably caused by temporary inactivity of the Sun's magnetic weather.

The grape harvest in wine-growing districts, such as Le Lavandou in the south of France, depends on good weather conditions. The

quality of wine vintages may follow the eleven-year cycle of the Sun's magnetism, through its influence on terrestrial weather.

follow the Sun's cycle. The link here is the Earth's weather. During sunspot maximum, there is generally better weather, and hence a better vintage and fewer cricket matches rained off. (The familiar sporting analogy is British because normally it rains quite a lot in Britain!) Although the correlation between solar 'weather' and terrestrial weather certainly seems to be real, meteorologists are hard pressed to explain it. It is easy to understand how energetic particles from active regions can affect the tenuous, electrically charged ionosphere of the Earth, but the total effect on the lower atmosphere should be minute in comparison with the energy it acquires directly from the Sun's heat – and the total energy output of the Sun varies very little during the cycle.

The Sun's weather affects our climate over a longer term, as well. In the late seventeenth century, all activity on the Sun's surface stopped for sixty years – and the Earth's temperature dropped. Only by 0·6°C (1·1°F) on average, it's true, but this was enough to cause a 'mini' Ice Age: the Thames in London, for example, froze over ten times in the severe winters.

Dr John Eddy, of the High Altitude Observatory, Colorado, has peered even further back into the past history of the Sun. By comparing mediaeval records of phenomena related to solar activity with the amount of radioactive carbon (carbon-14) in trees growing at the time, he has established that carbon-14 was more abundant during long periods of 'quiet Sun', like the late seventeenth century. This result caused no great stir, since carbon-14 is produced by the impact of cosmic rays from deep in space on atoms of the Earth's atmosphere; and when solar activity is weak, more of these fast particles can penetrate into the solar system.

In 1977, however, Eddy surprised the scientific world by

announcing that a study of carbon-14 in wood up to 3,000 years old revealed that for most of that time the Sun's surface has been very quiet. Periods of regular solar activity, like the present regime which has lasted since about AD 1715, occur only sporadically and rarely. For 90% of the last 3,000 years the Sun's face has been smooth and unblemished, very much the simple gas ball of the star theorists. It is only by chance that we are living at a time when the magnetic fields near the Sun's surface are letting loose their celestial firework displays.

The missing neutrinos

Looking back further still, the occurrence of the great Ice Ages, when ice sheets advanced down from the poles over the temperate regions of the Earth, may have been a result of events deep in the Sun's core. Many explanations have been proposed for the Ice Ages, and the most widely accepted astronomical theory at present involves slow changes in the direction of the Earth's magnetic poles, and in the shape of its orbit. But a rival hypothesis relates the intermittent freezing of the Earth's surface with the nuclear furnace at the heart of the Sun; and this has recently been seen in a new light as a result of the neutrino experiment described at the beginning of this chapter. Davis' first results, announced in 1968, caused consternation among astronomers: the number of neutrinos from the Sun was far smaller than expected. In consequence, the temperature of the Sun's core must be rather lower than the well-established theory of hydrogen-fusion stars predicted.

The initial panic caused by this announcement has now subsided, as astronomers have followed up this new clue to the Sun's workings. Years of collected results from the big underground tank have now given a fairly exact value for the neutrino flux, somewhat larger than the astonishingly low initial value; and theorists have recalculated their models of the Sun's interior, and now predict a rather smaller number of neutrinos. But a real discrepancy still exists: the Sun's heart is cooler than expected.

Astronomers are never wanting for an explanation, however, and in this case there are at least two hypotheses which can explain the difference between the real Sun and the theorists' model Sun in terms of simplifications in the model, without requiring a revolution in known physical laws. The first challenges the star theorists' long-held assumption that the original elemental mix of the Sun, including its core, was the same as its surface is today. If the present surface is actually composed of gas swept up from interstellar clouds since the Sun's birth, it is not necessarily a good guide to the original core material, and it is quite possible for the core to be cooler, in line with the neutrino results. And secondly, we must not forget that Davis' neutrinos left the heart of the Sun only eight minutes ago, speeding through the Sun and interplanetary space at the speed of light. The light and heat from the Sun's surface, however, has taken millions of years to travel up from its core, and so the Sun's present luminosity reveals how much energy was being produced at the core some millions of years ago. The Sun's nuclear reactor could vary periodically in temperature, by occasionally mixing in surrounding gas. If it has cooled slightly over the past few million years, the temperature deduced from the neutrino experiment should indeed be less than that predicted from the Sun's luminosity. Moreover, the present lower core temperature will cause a slightly cooler surface – and a cooler Earth – in a few million years' time. If this hypothesis is correct, we have a long-range forecast of a major Ice Age.

The Sun has been a bountiful supplier of warmth and light to the Earth for nearly 5,000 million years. The balance between nuclear reactions and gravity in the Sun's centre has kept its output remarkably constant over this very long time scale, for although slight changes may have caused our Ice Ages, the Sun's heat has never varied enough to freeze the Earth completely nor to boil away its oceans. To some extent there has been a fortunate coincidence. Theories of star evolution show that the Sun is gradually brightening, and is now shining about half as brightly again as it was in the early history of the solar system; but at the same time, the Earth's atmosphere has been depleted of

carbon dioxide by green plants, and as a result the 'greenhouse effect' (by which a carbon dioxide atmosphere keeps a planet hotter than expected) has decreased. The combination of these two factors has meant that the Earth's surface temperature has stayed between the boiling point and the freezing point of water. And we certainly have reason to be grateful, for only on so watery a world could our kind of life evolve.

But what of the Sun's distant future? Clearly our star cannot shine forever, because it is continuously depleting its stock of hydrogen 'fuel', and becoming choked with helium 'ash'. The nuclear fusion reactions which power the Sun can only occur right at its very centre, where the temperature and pressure are high enough, and calculations show that about one-tenth of the Sun's matter is in this central region. So, once the central tenth of the Sun's hydrogen has been 'transmuted' by fusion reactions into an inert helium core, our star will be in trouble. With no hydrogen in suitable conditions for reaction, the energy supply will dry up, and the Sun will have no support against the ever-present inward gravitational pull of its own mass.

In any case, we won't be around to witness the Sun's energy crisis. Theorists estimate that the Sun is roughly half-way through its hydrogen-fusion life, and the crisis will come in some 5,000 million years time. But carefully calculated mathematical models of stars, 'built' by theorists in computers, can be made to evolve as fast as we like. We can take a model Sun, and follow its progress after the central hydrogen is exhausted.

During its present hydrogen-fusion phase, the helium ashes accumulate at the very centre of the Sun, forming a dead core in the middle of the fusion region. One result of this build-up of 'clinker' is the gradually increasing brightness of the Sun; but not until the helium core has grown to one-tenth of the Sun's mass will the effects become obvious. The computer model shows that when the Sun's nuclear fusion reactor becomes choked, gravity will indeed begin to shrink the core. But strangely enough, the outer layers of the Sun will expand. This paradox is difficult to explain in simple terms, but the calculations leave us in no doubt that as most of the Sun's mass settles in closer and closer to the centre, there will be a reciprocal expansion of its tenuous outer envelope.

Observational astronomers can support their colleagues' calculations, for the sky is littered with giant stars whose properties show that they are beyond the stage of hydrogen fusion. Unlike the hydrogen-fusion stars, which form a neat sequence when their properties are plotted on a Hertzsprung-Russell diagram, the giants are a motley crowd. A hydrogen-fusion star with the same surface temperature as the Sun, for example, must have almost exactly the same brightness, the same mass, and the same size – identical in

131

all respects. Yet if we look at giant stars with the Sun's surface temperature, we find no set pattern: some are ten times larger and a hundred times brighter; others larger still, and correspondingly more luminous, right up to the enormously bloated *supergiant* stars shining ten thousand times brighter.

Most giant stars have cooler surfaces than the Sun: these are usually called *red giants*. Right out at the surface of such a star, gravity is comparatively weak and the gas there can contract and expand, changing the star's brightness as its size varies. The most regular 'breathers' are the Cepheid variable stars, yellow-hot supergiants so bright that they can be seen in other galaxies, and uniquely important because their pulsation period is accurately related to their average brightness. Once a Cepheid is identified by its regular – roughly weekly – brightening and fading cycle, we know what its intrinsic brightness is; and by comparing this with its apparent brightness, the distance can be easily calculated. Many giant and supergiant stars pulsate, though most do so rather more irregularly. Even the most familiar red giant of all, Betelgeuse (in the shoulder of the constellation Orion), changes its luminosity: over the course of a year, it brightens and then fades to only half or one-third of its maximum brightness.

The diversity among giant stars stems from the fact that under their similar-looking, pulsating surfaces, they have very different cores. All have left the hydrogen-fusion stage behind, and inside their extended, tenuous outer envelopes they have tiny, very dense cores, packed tight by gravity. The reactions and elemental make-up of the core depend on the star's mass and its age; and these are the factors that decide its size, surface temperature and brightness. So a giant star with a red-hot, 4,000°C (7,200°F) surface, shining 30 times brighter than the Sun, could be either a star 25% heavier than the Sun, 1,000 million years after its central hydrogen supply has been exhausted, or alternatively a star of twice this mass which passed the 'energy crisis' only 30 million years ago.

And so the giant stars remain inscrutable, their enormous envelopes concealing the state of their cores. Unlike hydrogen-fusion stars, it's difficult to tell anything about the age, the mass or the interior of any particular red giant or supergiant from studying its light output. We must rely on the theorists' computer calculations to guide us. These predictions – of the types of giant star, and the relative numbers we expect to see – agree satisfactorily with observations of real stars. And so let us return to the Sun's future, confident that our fast-running computer model is a reasonably reliable guide.

When its core becomes a shrinking helium clinker, and its outer envelope begins to grow, there is still a narrow shell around the core where hydrogen fusion can go on. So there is still energy for the Sun to radiate; and the shrinking of the core under the influence of gravitation also creates heat. In fact, when the Sun reaches its 'energy crisis', according to calculation, its total energy output will increase.

The increasing brightness of the expanding Sun will bode ill for the inner planets. Eventually, within 6,000 million years from now, the Sun's heat will bake the Earth's surface to unbearable temperatures, and will boil away the oceans. Any surviving, heavily protected human colony would see the Sun rising over the Earth's parched surface as a huge red ball, filling one-tenth of the sky. Its surface will gradually expand to encompass the orbit of Mercury, and that tiny planet will be boiled away as vapour in the giant Sun's outer layers.

But at this stage, the helium core will have shrunk so much that gravitational inpull will heat its centre to 1,000,000,000°C. Now the helium nuclei there will be colliding at such high speeds that they can combine in threes to make nuclei of carbon – and this helium fusion will suddenly provide a new energy source at the Sun's heart. The *core* will now expand slightly; and this will be counterbalanced by a sudden shrinking of the Sun's outer layers: our hypothetical watchers on Earth would see the Sun collapse – probably within a day – to a globe ten times smaller. But the gradual conversion of helium to carbon at the centre will lead to resumed growth of the Sun's outer envelope, and the red giant Sun will again begin to threaten the planets.

The Sun's interior has now adopted an onion-like structure. The outer layers are still unchanged since the Sun's formation, and consist mainly of hydrogen. Further in, there is a thin shell where the hydrogen is hot enough to fuse into helium, and within this, another shell of inert helium. Right at the centre is the new reaction region, where the helium is hot enough to fuse into carbon; and from here energy percolates upward to support the Sun against its inpulling gravitation.

And at this late stage the calculations become uncertain. The broad outlines are clear, but it's difficult to predict exact quantities. The Sun will probably grow to its maximum size in less than fifty million years from the beginning of helium fusion, and latest calculations show that it will not become much larger than its previous giant state. The Sun's surface, bigger than the orbit of now engulfed and non-existent Mercury, may threaten Venus; but the Earth's larger orbit should keep it out of harm's way. Our planet will again be baked dry, and any surviving human colony will now have another menace to face: the Sun's outer layers will become unstable, and will puff out into space as great shells of gas, each carrying away a small percentage of the Sun's matter. Over tens of thousands of years, half the gas in the Sun may be lost in this way, leaving the tiny hot

Ice Ages on the Earth, when ice sheets march down from the poles to normally temperate zones, may result from periodic changes in the nuclear reactions at the Sun's core.

core bare in its travel through space.

Astronomers are well acquainted with these expanding gas shells which surround some small, hot stars. They appear as tiny discs of light in a small telescope, like dim, distant planets; and they were rather misleadingly named *planetary nebulae* by the great early-nineteenth-century astronomer Sir William Herschel. Originally a musician, Herschel had an amateur interest in astronomy which led him to the accidental discovery of the planet Uranus and from then on he was supported by a pension from King George III. Together with his sister Caroline, he observed the sky on every clear night for forty years; and one of his great projects was classifying and measuring the positions of the various 'fuzzy' – nebulous – objects in the sky. Over a thousand of Herschel's 'planetary' nebulae are now known, and modern astronomers interpret them as evidence

of the final throes of stars like our own Sun, ending their lives, not with a large bang, but with a vague whimper.

White dwarfs

The central core of the Sun will then be effectively dead. The helium supply will have changed to carbon, and the temperature will never rise high enough for the carbon to fuse into other elements. The Sun's central skeleton, stripped of its outer covering, will just cool over the thousands of millions of years to come, releasing its stored-up heat into the cold of space. First white-hot, then yellow, cooling to red and finally fading to invisibility, our star will end up as an unseeable small relic in a vast Galaxy, still circled by a retinue of planets, including our now deep-frozen Earth.

Such skeleton stars, shining white, yellow or red, can be seen through a powerful telescope. They are small – no larger than the Earth – and so they shine only feebly. Astronomers have aptly named them 'white dwarfs'. The brightest star in the sky, Sirius, has a white dwarf companion, suspected long before its discovery because Sirius was found to move around a 50-year orbit; since Sirius is the Dog Star, its tiny companion is sometimes nicknamed the 'Pup'.

But what of the inexorable inpull of gravitation? Any body in space must be stressed inside by some force to oppose its tendency to collapse on itself. A planet is kept up by the interactions between the electrons in neighbouring atoms; in an actively shining star the high temperatures break up the atoms into nuclei and electrons, and it supports its outer layers by the rapid motion of these particles, heated by the nuclear reactions at its active centre. When the Sun's core becomes entirely composed of inert carbon nuclei, another type of pressure becomes important: the pressure of electrons themselves.

In Chapter 6 it was seen that, although electrons are infinitesimally small, the strange, dual wave-particle personality of matter means that two electrons cannot approach closer than about a wavelength; and that the wavelength decreases as the relative speed increases. In the Sun's contracting core, electrons and nuclei are moving about independently in the high-temperature plasma. But as the core grows smaller, the electrons eventually find that the space between them shrinks down to around one wavelength – and they begin to jostle for room. The only solution is for some to move faster, contracting their wavelengths and so taking up less room. In effect the gravitational force shrinking the star squeezes its electrons together until some must move faster; and then the jostling together and the fast-moving electrons provide pressure to hold the star up. Gravitation can thus be balanced.

The nuclei in the star are not affected by this great

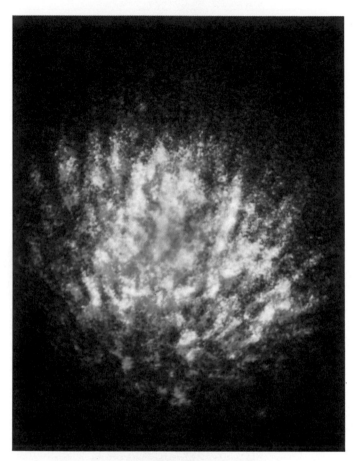

let the star assume a smaller size.

But there is a limit to what the electrons can take. When the fastest electrons are having to move at nearly the speed of light, it becomes more and more difficult for them to support the star: detailed calculations show that a white dwarf more than 45% heavier than the Sun just cannot support itself. If the core of a dying star is above this natural weight limit, its electrons cannot stop it from contracting indefinitely under the force of its own gravitation. Shrinking inexorably past the white-dwarf stage, it will reach one of the even more bizarre states which are known to modern astrophysics. Let us follow the history of a heavyweight star which will suffer this fate.

A heavyweight's demise

A star ten times heavier than the Sun is a spendthrift. We have already seen that it will spend only one-thousandth as long at its hydrogen-fusion stage, and it zips through the later giant stages at a correspondingly fast rate.

Its core will not stop at carbon, either. The greater gravitational pull will heat the centre until the carbon nuclei there fuse to heavier nuclei, such as silicon. Later, this 'inert' silicon core, contracting under gravitation, will heat up, and will then produce nuclei of iron. At this supergiant stage, the star's structure is even more 'onion-like' than that of the Sun as a giant star. Five different layers can be distinguished: a core of iron; successive shells of silicon, carbon and helium and their related elements; and an enormous outer envelope consisting mainly of hydrogen. And between the layers there are fusion 'shells', where nuclei of each higher layer are fusing and dropping down to the lower. The star's total energy output comes from many different reactions at different depths within it.

But the succession of 'fuel' to 'ash', to be used again as 'fuel' at a higher temperature, cannot last forever. Very large nuclei are unstable because of the electrical repulsion between the protons in them, and to fuse iron nuclei together to make heavier elements is an uphill task: it doesn't produce any energy, but needs to take energy in. The iron core thus faces a crisis quite unlike that suffered by the star before: there is now no way in which it can produce an energy outflow. And gravitation is inexorably pulling in the star around it, irrevocably squeezing it smaller and smaller, hotter and hotter.

Something must give, and the iron core eventually gives up the unequal struggle. The increasing temperature means that the individual nuclei are moving faster; and their collisions become more violent, until they literally smash each other apart. The fragments are helium nuclei; but there is a price to pay. During the star's supergiant life, it has painstakingly fused helium nuclei into successively larger nuclei, radiating away the excess energy coming from these reac-

crush. Since the wavelength of a piece of matter depends inversely on its mass, and the nuclei are thousands of times heavier than the electrons, they have plenty of elbow room and move about quite unimpeded. Theoretically, it's possible to have carbon white dwarfs, iron white dwarfs or helium white dwarfs: in each case the nuclei – so important in planets and stars – do nothing but provide positive charges to cancel overall the electric charge of the electrons doing the 'donkey work'. This state is known as *degenerate* matter (with the result that some technical papers have titles which must seem strange to the non-astronomer: American astrophysicist Jeremiah Ostriker, for example, has written a review of 'recent developments in the theory of degenerate dwarfs'!).

The tight packing in a white dwarf is beyond comprehension: a typical white dwarf contains about two-thirds as much matter as the Sun, but compressed into a sphere as small as the Earth. It's matter a million times denser than water: a thimbleful of white-dwarf material would weigh a ton. And because of the peculiar nature of the force supporting white dwarfs, the more massive stars are smaller than the lighter ones. More mass means a greater gravitational pull and more electrons forced into the faster, shorter-wavelength state. These can pack more closely, and

*The Sun will eventually swell
to become a red giant star, a
hundred times its present size.
The innermost planet, Mercury,*

*will evaporate as it is immersed
in the Sun's outer layers, and
the Earth' surface will be baked
dry.*

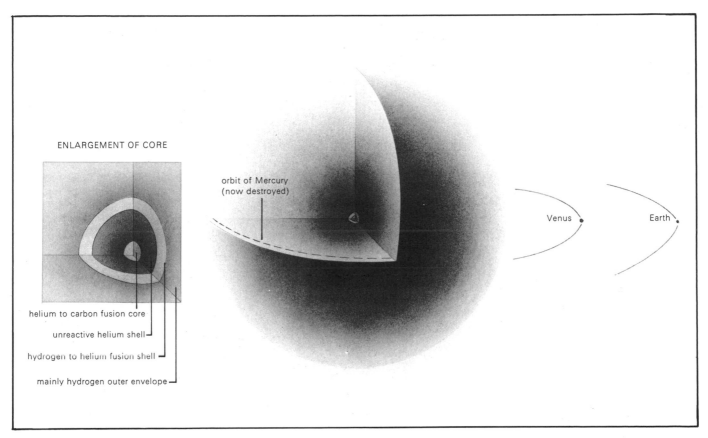

ENLARGEMENT OF CORE

helium to carbon fusion core

unreactive helium shell

hydrogen to helium fusion shell

mainly hydrogen outer envelope

orbit of Mercury
(now destroyed)

Venus

Earth

tions. When the iron nuclei smash up, they must restore this lost energy; and this can only come from the motion of the nuclei. The resulting helium nuclei thus have less energy of motion than the iron nuclei, and they can't support the core as well. Gravitation squeezes the core inwards even more drastically; the temperature rises faster; more iron nuclei disintegrate; and as a result the core loses support and collapses faster. The end has come: the core plunges towards headlong collapse.

The star's mortal throes have so far been well concealed from the outside Universe by its huge gaseous envelope. But the final collapse does not remain secret for long. The suicidal smashing of nuclei produces a vast flood of neutrinos from the star's core, and an astronomer on a planet near a supergiant star could diagnose the onset of the star's terminal illness by studying the flow. But most of the final flood of neutrinos is absorbed by the intermediate layers of the star itself; the energy they carry lifts off and blasts out the outer layers of the star in a tremendous explosion known to astronomers as a type II supernova.

Here we have a prime example of the interplay of particle physics and astrophysics at the borders of present knowledge. The first calculations that were made indicated that most of the neutrinos would stream through the star, without affecting the outer layers, and that the star's envelope

would just be pulled in by gravitation. Yet astronomers knew that supergiant stars explode, and neutrinos from the centre seemed to be the only possible cause. The inconsistency now seems to be resolved, for the 'neutral current' reactions predicted by the combined electromagnetic-weak force theory of Weinberg and Salam should produce a sufficient force to stop the neutrinos within the star. Theorists are still working on the detailed calculations, but it's probable that the astronomer's embarrassment at explaining the cosmic violence of type II supernova explosions can be relieved by the latest theories and discoveries in the realm of the fundamental particles.

The enormous shell of gas exploded into space contains as much matter as several Suns, and it is expanding outwards at some five thousand kilometres (3,000 miles) per *second*. It will continue to expand for perhaps half a million years, sweeping up the thin gas between the stars all the while. Its gradually increasing bulk slows down the expanding shell (the supernova *remnant*) until it eventually merges in with the general interstellar gas as a ring of denser clouds.

Astronomers have only recently begun to study supernova remnants in detail, for only a few emit much visible light. Radio astronomers can easily pick up their signals, however; and the latest catalogues contain over a hundred of these tattered ghosts of dead stars. The young science of

135

Sunrise, AD 6,000,000,000. The red giant Sun rises over a scorched and desolated Earth.

X-ray astronomy promises to reveal more, for the gas in the shell shines brightly at these wavelengths, owing to its very high temperature of millions of degrees. The immense speed of the expanding shell heats up the swept-up gas, just as the air caught up at the nose of a supersonic plane is heated by the plane's speed. Supernova remnants are of vital importance in our Galaxy, for a massive star is a celestial phoenix: from its expanding 'ashes' are born new generations of stars.

We have come full circle now. In looking at star birth – and planet birth – we found that a gas cloud in space could be compressed by an expanding supernova remnant; some of the stars born in this cluster will be heavy stars, which, within 10 million years, will explode; their expanding remnants will in turn compress existing gas couds further away and trigger off a new generation of stars. By returning their gases to the interstellar 'pool', supernovae return their debt – and with interest, for they enrich the Galaxy in

'heavy' elements. Their exploded shells carry with them elements like carbon and silicon which have been formed in the star's fiery interior, and scatter them through space.

In the beginning, it is now believed, our Galaxy consisted only of hydrogen and helium gas. All the other elements were made deep inside stars, and were later spread out, mainly by supernova explosions. In this way, later-generation stars, like the Sun, contain some 2% of heavy elements; and, more important to us, the forming Sun was surrounded by dust made up of these elements, which became its retinue of planets. Apart from hydrogen and helium, all the nuclei in the atoms composing the rocks of Earth, the air around us, and our bodies themselves were synthesized in the multi-million degree nuclear furnaces within the overweight stars.

Here again, Nature's forces have been well deployed. Gravitation has taken the star past its hydrogen-fusion stage, increasing its core to successively higher temperatures. At the higher speeds so achieved, the successively heavier and more positively charged nuclei have been able to approach more closely as they 'bounce' apart by electrical repulsion; and so the short-range strong force has had the chance to fuse them into still larger nuclei. Eventually, at the iron-core stage, the balance of strong and electromagnetic forces within each nucleus decrees that larger nuclei will be unstable, and gravity takes hold again to crush the core. The break-up of nuclei results in a weak-force reaction which generates a flood of neutrinos, and again by a weak-force 'neutral-current' reaction these blow off the star's outer layers and distribute the large nuclei through space. It's ironic that the type II supernova explosion – a cataclysm in which a star temporarily becomes a hundred million times brighter than the Sun – is a product of the so-called 'weak' nuclear force!

Pulsars

What of the core of the supergiant star, which we left collapsing faster and faster after it had given up the unequal battle against gravity? The more daring theorists had suggested possible answers back in the 1930s, when the theory of star evolution was in its infancy; but the experimental confirmation of their seemingly wild ideas had to await the new astronomy of the 1960s and '70s.

Near Cambridge, England, in the autumn of 1967, Antony Hewish and his research students built a new type of radio telescope, intended to probe the fine detail of small radio sources way beyond our own Galaxy. This revolutionary telescope is not the usual dish: too large to be steerable, it resembles a field full of wire clothes-lines. It uses, as a vital first component, the stream of electrically charged particles escaping from the Sun: the 'solar wind'. As this whistles out past the planets, tiny irregularities in it affect the radio waves coming in from the depths of space, and to the radio astronomer the source seems to 'twinkle'. This is actually a good word to use, for the effect is very similar to the way that small irregularities in the Earth's atmosphere make the light from stars twinkle: Hewish, indeed, jokes that the optical astronomers did not entirely understand the twinkling of starlight until radio astronomers did the calculations for radio waves in the solar wind.

The amount of twinkling we see depends on the apparent

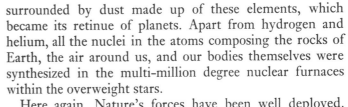

The Helix Nebula in Aquarius is a shell of gas puffed off from the star right in the centre. (The other stars are simply in the background or the foreground.) Such nebulae resemble the discs   *of planets when seen in small telescopes, and they have been given the misleading name 'planetary nebulae'. Photograph from the Mount Wilson and Palomar Observatories.*

size of the distant source (that's why planets twinkle less than stars), and Hewish's telescope was designed to pick up details of the twinkling. But simply because it was the first sensitive radio telescope to react to quick changes in the brightness of a radio source, the Cambridge instrument came up with some completely unexpected results.

Among the reams of paper chart recording the 'twinkling' sources, one of Hewish's students, Miss Jocelyn Bell (now Dr Jocelyn Burnell) noticed one which did not fit the pattern. A closer look revealed why. Instead of twinkling erratically, this source was flashing regularly, once every 1·337 seconds. It could easily have been man-made interference; yet their continued observations showed that it appeared in the record four minutes earlier each day. This would be strange for a terrestrial source; but would fit in with the fact that a source in the sky will drift into the telescope's fixed beam four minutes earlier each day, as a result of the Earth's movement around the Sun. The only possible interference could be from equipment running at this faster 'sidereal' rate in another astronomical observatory, and switching at 1·337 second intervals; but queries to these observatories quickly ruled this out. Some powerful radio source in the sky was 'pulsing' – and it was dubbed a *pulsar*.

Speculative minds might at once have concluded that the 'four-acre telescope' was picking up signals from a cosmic beacon, a communication from an extra terrestrial civilization. It was important to check whether such signals could arise from a natural source before the sensational newspapers heralded the discovery of extraterrestrial life, so Hewish's team carried out further investigations before releasing the news. The pulsar signal was so precise that they could measure the very slight change in the length of its period as the Earth orbited the Sun; and since the

A supergiant star is the final stage in a massive star's life; nuclear reactions have created concentric shells of different elements. The iron core will

eventually collapse, releasing a flood of neutrinos which blow off the outer layers in a supernova explosion.

*Below:
The gases thrown out from a supernova sweep up the interstellar gas and heat it to around 10,000,000°C. Such gas emits X-rays copiously, and this false colour picture*

represents the X-ray emission from Cassiopeia A, the 300 year-old remains of a supernova (other pictures on p 186–187). Based on Copernicus Satellite data obtained by Charles, Culhane and Borken.

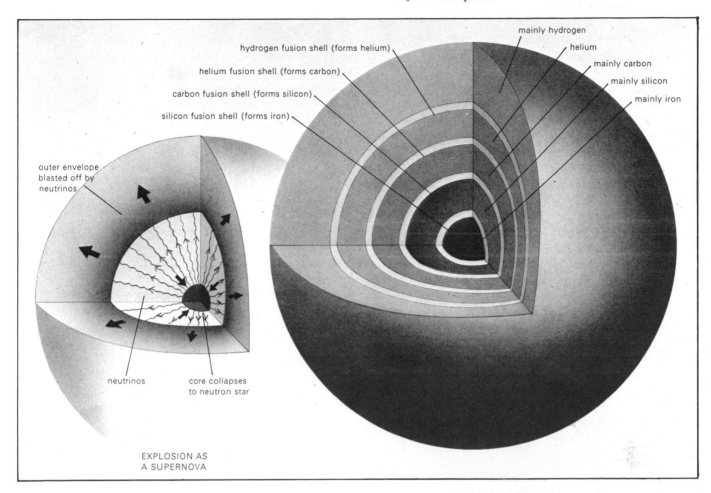

hydrogen fusion shell (forms helium)

helium fusion shell (forms carbon)

carbon fusion shell (forms silicon)

silicon fusion shell (forms iron)

outer envelope blasted off by neutrinos

mainly hydrogen

helium

mainly carbon

mainly silicon

mainly iron

neutrinos

core collapses to neutron star

EXPLOSION AS A SUPERNOVA

change fitted expectation exactly, the radio source itself could not be moving in orbit around another star: it was not a beacon fixed to a planet. This seemed to rule out an interstellar message; and Jocelyn Bell clinched the case for a natural origin when she searched the records and discovered three more, weaker pulsars. It was too much to believe that four different civilizations would be trying to contact Earth simultaneously.

As observations improved, it became obvious that the great regularity of the pulses could only be caused by the spinning of a very small and dense star – even smaller than a white dwarf. Only one candidate fitted the bill: the neutron star. In 1932, less than a year after the neutron was discovered, the brilliant Russian physicist Lev Landau had predicted that the Galaxy may contain stars made entirely of neutrons. Just as white dwarfs are held up by the pressure of the crushed-together electrons, a neutron star is held up by the jostling of neutrons packed so tightly that their separation is roughly one neutron wavelength. Since neutrons are almost two thousand times heavier than electrons, their wavelengths are smaller and they can pack

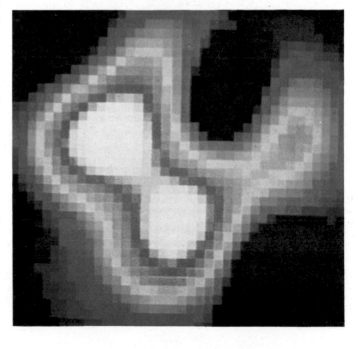

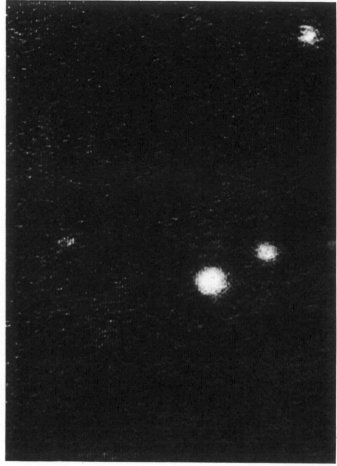

that much closer together. A neutron star is only a few kilometres across, yet contains as much matter as the Sun: a pinhead of matter from a neutron star would weigh a million tons.

In 1933, two American astronomers, Walter Baade and Fritz Zwicky, pointed out that neutron stars could form from the imploded cores of supernovae. As the core collapses past the white-dwarf stage, the electrons in it are forced closer together than they are 'allowed', and their only remedy is to combine with protons to form neutrons. In this extreme case, neutrons are the most stable particles, for a decay into proton and electron is ruled out by the fact that there just isn't any room for such an electron to occupy.

For over forty years neutron stars were relegated to a backwater of astrophysics. Even if they existed, astronomers thought, they would be too faint to detect. With the discovery of pulsars, however, the old theories were brought out and dusted off; and a rotating neutron star had to be the answer to the pulsar mystery. Such stars have strong magnetic fields, compressed because they are trapped in the collapsing core, and somewhere in this field (the exact details are still in dispute) a beam of radio waves is pro-

duced. The spinning star sweeps this beam around the sky like the revolving lens in a lighthouse; and the radio astronomer on Earth sees the 'flashes' or 'pulses' of this radio signal just as a mariner sees the regular sequence of flashes from the lighthouse.

Final vindication came in 1968, when radio astronomers discovered a pulsar in the famous Crab Nebula. This strange nebulosity had been christened in the nineteenth century by the third Earl of Rosse, who discovered claw-like filaments extending from the oval nebula. He was observing with a huge and unwieldy telescope at his ancestral home in central Ireland; surprisingly enough, his mammoth telescope remained the largest in the world for 72 years, until the '100-inch' Hooker telescope was opened in 1917 at Mount Wilson, under the clear Californian skies. Modern photographs show the contorted nebula better than the eye can see at any telescope, and the mystery of its strange appearance was to some extent explained when it was found to coincide with the position of a supernova seen and recorded by the Chinese in AD 1054 – a star explosion so bright that it was visible in daylight for 23 days. Yet the Crab Nebula is an unusually powerful super-

The Crab Nebula is the remains of a star seen to explode by the Chinese in AD 1054. The outer layers have formed twisted filaments which led to the nebula's name, while the original star's core has collapsed to become a neutron star, or pulsar (see opposite).

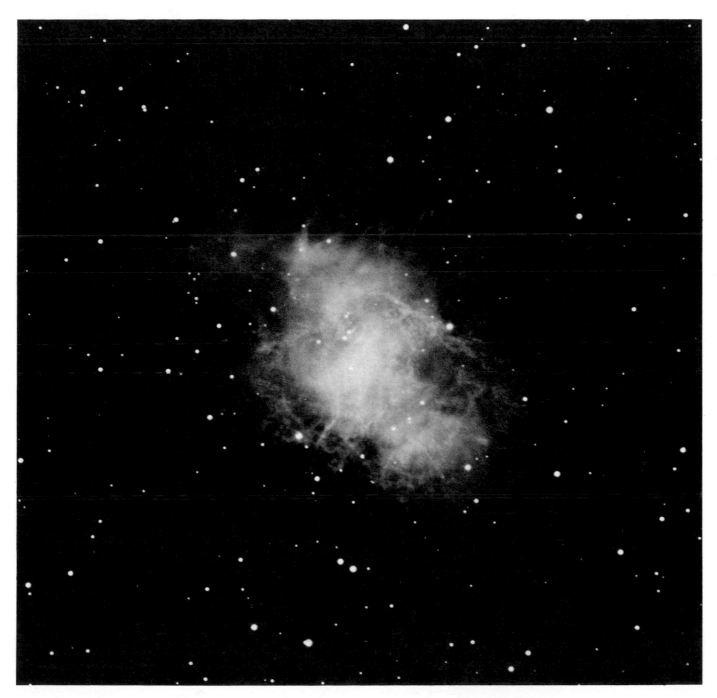

nova remnant, both as a source of light and of radio waves.

The discovery of the Crab pulsar solved several mysteries at once. Its position in the centre of the nebula showed that supernova cores can indeed become neutron stars; and it matched the position of a faint visible star which had long been suspected to be the dead star's core. The Crab pulsar is the fastest known, flashing 30 times a second. (A close-up view of it defies the imagination: a ball 25 kilometres (15 miles) across, weighing about as much as the Sun, and whirling around 30 times a second.) Optical astronomers immediately cracked down on the visible star and, using special techniques, discovered that it, too, flashes at this rate: too fast for the eye to see, and with such a short period that its light is averaged out on a photograph.

Further research has shown that all pulsars are very gradually slowing down as they lose rotational energy. The Crab pulsar is no exception; and its lost energy is conveyed to the surrounding Crab Nebula: this is the secret of the

The radio waves from pulsars – and light from the Crab and Vela pulsars – must come from fast electrons moving in the pulsar's magnetic field, and

emitted as a narrow 'lighthouse beam'. Two possible sites for the radio emission are shown here – only further research can decide which alternative is correct.

nebula's unusual brightness.

These observations of pulsars dovetail neatly with the theory of neutron-star interiors. The bulk of the star is made of the degenerate neutrons, moving completely freely (as a *superfluid*), and currents of moving neutrons can readily generate the pulsar's strong magnetic field. Right at the centre, neutrons are so crushed that they can react, and produce a small proportion of heavier assemblages of quarks: the 'old fundamental particles', named after the Greek letters sigma, lambda and delta. The neutron star's core has been aptly called a 'rich Greek alphabet soup'.

The outer layer of the neutron star is effectively a solid crust, constructed like a white dwarf, with iron nuclei spaced out in a 'sea' of degenerate electrons. It may have mountains on the surface, but the neutron star's powerful gravitation would prevent these from being more than a few millimetres high; nevertheless, the strong pull would also mean that it would take more energy to climb a one-millimetre bump on a neutron star than to climb Earth's Mount Everest! The solid crust can suffer 'starquakes' too, as the pulsar slows down and its equatorial bulge – caused by the high spin rate – tries to decrease. Radio astronomers see these as sudden increases in the rotation speed, and they have observed a sudden speed-up of the Crab pulsar which was due to the equatorial radius suddenly shrinking by a

mere one-hundredth of a millimetre.

Pulsars are no strangers to the radio astronomer now. In the past ten years over 300 have been found, with rotation periods ranging from the Crab pulsar's 1/30 second to just over 4 seconds. Yet only one more optically pulsing source has been found. Lying in the southern constellation Vela, this 'visible' pulsar was picked up by astronomers manning the large Anglo-Australian telescope at Siding Spring, New South Wales. They used electronic detectors designed to register only light pulsing at exactly the frequency of the radio pulsar, and they found a pulsating source too weak to measure by any other method: the press declared that British astronomers had found the faintest star ever detected.

Astronomers working out the statistics of pulsar periods and slow-down rates have calculated that an average pulsar broadcasts its radio waves for only a few million years. Then its pulses become more sporadic, and it eventually fades out. More surprisingly, the statistics show that pulsars must be born every five years or so in our Galaxy to keep up the observed numbers – yet supernovae occur far less frequently than this. If this conclusion is correct, perhaps only a small fraction of pulsars are the collapsed cores of supernovae, and astronomers will have to look to other types of star as the progenitors of most pulsars – which means pulsars may still have surprises in store.

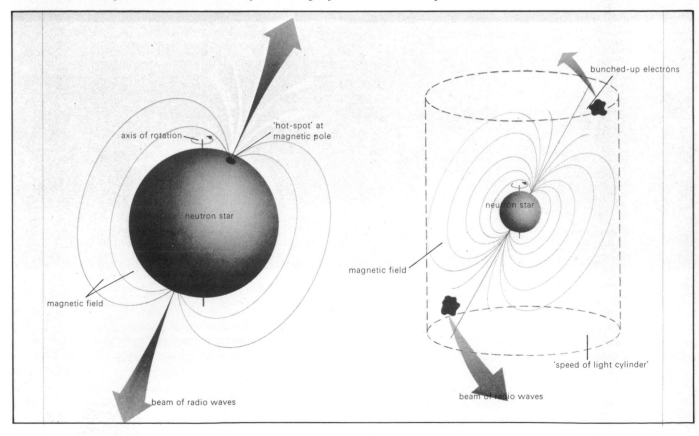

A star is a phoenix – from its ashes rise new stars. Here is the central region of the Orion Nebula, where the four innermost stars (the Trapezium) have recently formed. Many nebulae only begin to collapse to form stars when compressed by a nearby supernova explosion; and certainly the young Trapezium stars contain elements created in previous generations of now-disrupted stars. Lick Observatory photograph.

The Veil Nebula in Cygnus is the tattered remains of a 20,000 year old supernova explosion. The small core of the original star may have collapsed so utterly that it has become a black hole, a region of space with such strong gravitation that it is cut off from the rest of the Universe.

'Black hole = naughty word.' So went the message scrawled on the blackboard at an International Astronomical Union session in 1970. And indeed, practical astronomers were not alone in dismissing this unlikely offspring of the theorists' imagination, for a black hole really seems like science fiction: a small region of space where matter disappears forever from our Universe, as down some kind of nightmare 'hole in space'; and a hole which can't be seen because it absorbs light, too. No wonder many scientists were unprepared for the idea of black holes in our Universe.

Yet opinion has swung around. Astronomers have been led, by a process of elimination, to accepting black holes as the only way of explaining certain astronomical objects. Like Sherlock Holmes, they have had to allow that when the impossible has been eliminated, whatever remains – however improbable – must be the true answer. And as black holes have begun to win acceptance among astronomers, the bizarre nature of these objects has led to a rash of sensational popularizations: travel to other universes, 'mini' black holes colliding with the Earth, 'white' holes. What is the truth about black holes?

The idea, surprisingly enough, dates back to 1795, to Pierre-Simon Laplace's great treatise on gravitation and the solar system, *Exposition du système du Monde*. According to Newton's theory of gravitation, the strength of the inward gravitational pull at the surface of a spherical object depends on both its mass and its radius. If the Earth contained twice as much matter, but had the same radius as it does, we would all feel its gravity twice as strongly. Similarly, if the Earth were simply shrunk to a smaller radius, but with the same amount of matter as it has at present, the surface gravity would increase. We can also think of this in terms of *escape velocity*. Neglecting air resistance, anything thrown upwards faster than 11 kilometres (6·9 miles) per second can escape completely from Earth. Although it is continuously slowed down by Earth's gravity, the effect diminishes as it moves away, so that it is never brought to rest. Escape velocity, too, depends on the mass and radius of the body in question: Laplace realized that a body with the same density as the Earth, but 27,000 times larger, would contain 20 million million times as much matter, and with these conditions the escape velocity would be equal to the velocity of light (300,000 kilometres [186,000 miles] per second). He thought of light as travelling like a succession of small projectiles, and quite correctly deduced that light couldn't escape from such a body – in other words, the largest and most massive stars in the Universe could be invisible.

Although we now know that Laplace's hypothetical stars can't exist, his basic idea was sound. With a combination of high mass and small radius it's possible to have a body from which light cannot escape. But modern theorists think of black holes in rather different terms, for Newton's theory of gravitation has been supplanted by Einstein's, and the idea of light travelling as simple projectiles has been replaced by the wave-particle duality of quantum theory.

The celebrated patent clerk
Albert Einstein's great contributions to science began in 1905, when he was a twenty-six-year-old clerk at the Patent Office in Berne. He wrote three important scientific papers that year; and one of them was destined to alter our way of thinking about the concepts of *space* and *time*. Unassumingly titled *On the electrodynamics of moving bodies*, it was the first real plunge into the meaning of the combined electric and magnetic theory, welded together by Maxwell some forty years earlier. The nettle which no one until Einstein had grasped was that the theory predicted that the speed of light is always the same, to all observers. If you measured the speed of light coming from the Sun, it would turn out to be 300,000 kilometres per second. Now suppose you could blast off in a rocket sunwards at 200,000 kilometres per second. In the few minutes you would have before making a devastating rendezvous with the solar surface, would sunlight's speed seem to be 200,000 + 300,000 = 500,000 kilometres per second? Maxwell's equations say no: you would still measure it to be 300,000 kilometres per second! And there was no doubt that Maxwell's theory was basically correct, for it explained the properties of light, as well as electric and magnetic forces, and it had predicted radio waves before their discovery.

Einstein followed through the implications of this unexpected result, and discovered that it could not fit in with the intuitive idea that space and time are the same to everyone, no matter how they are moving. If two identical spaceships pass each other at a very high speed – say, 270,000 kilometres per second, 90% of the speed of light – the crew of each would be surprised by the appearance of the other ship. It would be only half as long as expected; and if its mass could be measured as it sped by, it would seem to be twice as massive. Moreover, comparing clocks on board, the clocks on the other ship would seem to be ticking only half as fast as those on one's own ship. Those effects are completely complementary: each ship's crew sees the other suffering these strange effects, and since it is the relative motion of the ships which is important – each crew experiences that its ship is at rest, and that the other is moving – the theory acquired its name of relativity.

Such notions as the length and mass of an object changing just because it is moving relative to us seem absurd in everyday life; but that is simply because all the speeds we are used to are interminably slow compared with the speed of light. A car passing by at 100 kilometres (62 miles) per hour is contracted by about the diameter of an atomic

Sir Isaac Newton first formulated a theory of universal gravitation, after he was inspired by seeing an apple fall. This centuries-old theory – now superseded by Einstein's – actually predicts the existence of black holes, usually regarded as a very modern idea.

nucleus – far too small to measure. In particle-physics accelerators, however, where subatomic particles are accelerated to almost the velocity of light, these effects are very noticeable. Although particle physicists can't measure the size contraction of point-like particles, they readily detect the increase in mass, and the slowing down in time-keeping – unstable particles live far longer when they are on the move.

With increasing speed, all the effects intensify. Approaching the speed of light, natural clocks slow until they become effectively frozen; lengths contract to zero; and mass increases indefinitely towards infinity. For this reason, it's impossible to achieve the speed of light, let alone surpass it. We would have to keep putting increasingly more energy into the more and more massive body, with decreasing returns in the way of speed. The speed of light is the natural speed limit of the Universe (light photons themselves travel at the speed limit, because they have effectively zero mass: you can't make a photon go *slower* than the speed of light).

Two years later, Einstein realized that because energy of motion increases a body's mass, there is an equivalence between mass and energy. As a result, matter at rest could be turned into pure energy, given the right circumstances.

This relationship can be summed up in the famous equation $E = mc^2$, verified now in nuclear reactors and, more infamously, in nuclear bombs.

Einstein vs. Newton

But this first theory, the Special Theory of Relativity, didn't include gravitation, nor the effects of acceleration. It was impossible simply to graft Newton's theory on to Special Relativity, and to produce a sensible gravitational theory. It took Einstein ten years to extend the Special Theory, and his General Theory of 1915 is still the accepted theory of gravitation today.

His starting point was a strange fact of nature. If you are driving, and put your foot down hard to accelerate, you are pushed back into your seat by inertia as the car speeds up. If you continue to accelerate smoothly, the continuous force pushing you back feels the same as the force of gravity which is pulling you downwards; Einstein accepted that there was some kind of basic equivalence of gravitational force and inertial force, and made this the pivot of his theory. In essence, according to Newton's theory, you could tell that the car was accelerating forwards, rather than the Earth and road backwards, by the inertial force pushing you into the seat. Einstein needed a theory which preserved the idea of relativity: that it isn't possible to define whether car or road is accelerating (note that if your speed is constant, you can't say which is moving, and it's a Special Relativity case).

Suppose we think of the car as stationary, and that the rest of the Universe is accelerating backwards. To explain the force you feel pulling you backwards, you have to say that when the whole Universe (apart from your car) accelerates, it produces an extra *gravitational* force which pulls you back. Although 'thought experiments' like these may seem hopelessly philosophical, from them Einstein was able – with the help of some intricate mathematics – to formulate a new theory of motion and gravitation. Not surprisingly, it predicted the same results as Newton's when gravitation is weak – for the orbit of the Moon around the Earth, for example – but in strong gravitational fields, there were striking differences.

His first success was with the orbit of Mercury. This small planet moves in an elongated path close to the Sun, and the long axis of the orbit gradually moves around. Most of this motion is due to the gravitational attraction of all the other planets, and could be explained by Newton's theory of gravitation; but the expected and measured rates did not quite match up. Astronomers searched in vain for a small planet which might circle the Sun closer than Mercury, and could be perturbing its motion. Mercury, close in where the Sun's gravitational pull is relatively strong, was the one planet for which Einstein's equations gave a measurably different answer from Newton's; and Einstein's agreed with the observations. Mercury's 'anomalous' motion was simply due to astronomers using the wrong gravitational theory.

A more convincing effect, because Einstein predicted the results before the observations were made, concerned the bending of starlight by gravity. When a star lies almost behind the Sun in the sky, its light has to pass through the intense region of the Sun's gravitational field; Einstein showed that its path would be deflected slightly, so that the star would appear to be in a slightly different position. Stars can only be seen near the Sun when the light of the latter is blocked out by the Moon during a total eclipse, and the first opportunity to test Einstein's prediction came with the end of the First World War. A British expedition on the small island of Principe off the African coast observed the eclipse of 29 May 1919 – and their photographs showed that the star images had been displaced by the amount that Einstein had predicted. The war-weary world seized on this timely example of the true internationality of science; and, without even really understanding his achievement in re-casting the structure of space and time, the popular press made Einstein's name a household word all over the world.

The third so-called 'test' of General Relativity is directly related to the subject of black holes. Einstein showed that light loses energy when it travels against a gravitational field. Since an individual light photon always travels at the same speed (the speed of light), the energy it loses in escaping has to come from the photon's internal energy. And according to Planck's formula relating energy and wavelength, the decreasing energy of the photon means that it appears as radiation of longer wavelength. In theory, then, a photon emitted from the Sun's surface will arrive as a photon of slightly longer wavelength (shifted slightly towards the red end of the spectrum) when we receive it on Earth. It's possible to measure the red shift, because atoms in the Sun's atmosphere emit light at specific wavelengths – the spectral lines. Comparing the wavelengths of the Sun's lines with those of the same kind of atoms in the laboratory, astronomers have found that the Sun's gravitation has indeed shifted its lines by Einstein's predicted amount. (More precise laboratory experiments, using gamma-rays travelling upwards in the Earth's gravitational field, have verified the Einstein shift even more accurately for these shorter-wavelength photons.)

Astronomically speaking, the Sun's gravitation is pretty weak: it red-shifts the spectral lines by only two parts per million. The Einstein effect on radiation is related to the massive body's escape velocity, however, since in either case a 'test particle' travelling upwards is losing energy struggling against the gravitational pull. A white dwarf star, with roughly the same mass as the Sun, but a radius as small as

the Earth's, has an Einstein shift a hundred times greater. Even so, it is not easy to measure accurately the wavelengths of lines in the spectrum of such faint stars; and there's an added problem, for a white dwarf's spectral lines will be displaced slightly by the Doppler effect, owing to its motion towards or away from us. By selecting white dwarfs in double-star systems, astronomers can alleviate part of the problem, because the average wavelength shift of the other star (due entirely to the Doppler effect), taken over a whole orbit, tells us the speed of the double star through space. The white dwarf is then found to have an average wavelength shift slightly redder than its companion, and to within the limits of this difficult and rather imprecise measurement, the results agree with the predicted Einstein shift.

Moving on to smaller and even denser objects, neutron stars claim our attention. Einstein's theory predicts that the red shift for a neutron star as massive as the Sun, and only 25 kilometres (15 miles) across, is almost 15%. A photon struggling up from the surface of such a neutron star has to give up one-seventh of its energy in its bid to escape from gravitation. Unfortunately, neutron star surfaces do not emit spectral lines, so this prediction can't be tested directly, but it does show that even among the now accepted members of our Galaxy, strong gravity can affect the way we see things.

Here we enter the realm of the black holes. Just as there is a limit to the maximum mass which a white dwarf can have before the electrons have to give in to gravitation, neutron stars, too, have a mass limit. When gravitational compression becomes too high, the neutrons react to form heavier particles – those of the 'Greek alphabet soup' mentioned earlier – whose shorter wavelengths allow the neutron star to shrink further. The mass limit depends on exactly how nuclear reactions take place at the high energies involved: once again, astrophysicists need the high-energy results of particle physicists to help them understand matter at its breaking point. But neutron stars more than three times heavier than the Sun certainly cannot support themselves against gravity (and the limit may be lower than this).

Again let us follow the fate of a heavyweight star, this time a star thirty times heavier than the Sun. Such a star loses a lot of matter from its surface as a 'stellar wind' when it becomes a supergiant, and it can blow off most of its outer layers in a supernova explosion. Yet its collapsing core will still be some five times heavier than the Sun. Collapsing past the white-dwarf state, for it is well over the weight limit for such stars, it shrinks to neutron-star densities. But again, it is above the limit for neutron support. Collapse continues. As the radius of the core contracts, the escape velocity at its surface increases more and more rapidly; and light photons have more difficulty in

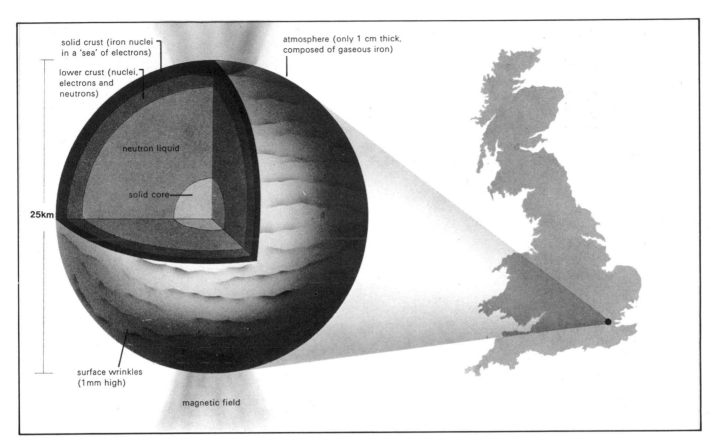

solid crust (iron nuclei in a 'sea' of electrons)

lower crust (nuclei, electrons and neutrons)

atmosphere (only 1 cm thick, composed of gaseous iron)

neutron liquid

solid core

25km

surface wrinkles (1mm high)

magnetic field

escaping from the gravitational field. Eventually – and this is all within one second – the gravitational pull at the surface of the star becomes so strong that the light photons cannot escape: they use up all their energy in the struggle, without succeeding.

From a safe vantage point outside the star, we would see – if we had a telescope powerful enough – the core shrink rapidly until it reached its critical size, and then it would suddenly fade out in a matter of a thousandth of a second. The tiny, fiercely glowing star core would have been cut off from our vision by its own gravity. In fact, once it shrinks beyond its critical size and we can no longer see it, we can know nothing more about the dead core. Light and other radiations cannot escape from its gravitational field. The critical size is, not surprisingly, the size at which the escape velocity has increased to the speed of light; and, since no material object can move faster than light, no particles of matter can escape from the collapsed core's gravitational field either. As the core grows smaller, we can never know of any events which go on within an imaginary sphere, equal to the core at its critical size. This imaginary surface, enclosing a small region of the Universe which we cannot know anything about, is called the *event horizon* – or more graphically, a *black hole*.

A black hole is 'black' since neither light nor any other radiation can escape from it. It is a 'hole', because objects can fall into it, but can never escape again. Even a beam of light shone at a black hole would be captured by it, and no light would be reflected back to us. Science-fiction writers have had a heyday with these invisible black holes in space, lying in wait for the unwary space traveller who has no idea of his fate until he notices – too late – that his ship is being pulled faster and faster by the gravitation of something invisible. Once he is pulled through the event horizon, his colleagues outside lose contact with him forever. The space traveller can still see the outside Universe, since light from outside can easily get into the hole, but his unique view of the inside of a black hole will be short-lived as he tumbles down to its centre.

In outline this scenario is true, but black holes have acquired a false reputation as greedy scavengers, 'sucking in' anything which comes within the outermost fringes of their gravitational fields. In fact, the gravitational pull of a black hole outside its event horizon is no different from that of any other celestial body. The Sun is pulling on the Earth all the time, and yet the Earth is not 'sucked in' to our star because its orbital motion exactly counterbalances the Sun's pull. If the Sun suddenly shrank to a black hole, the gravitational pull felt by the Earth would not alter in the slightest: our planet would continue to circle its new, dark

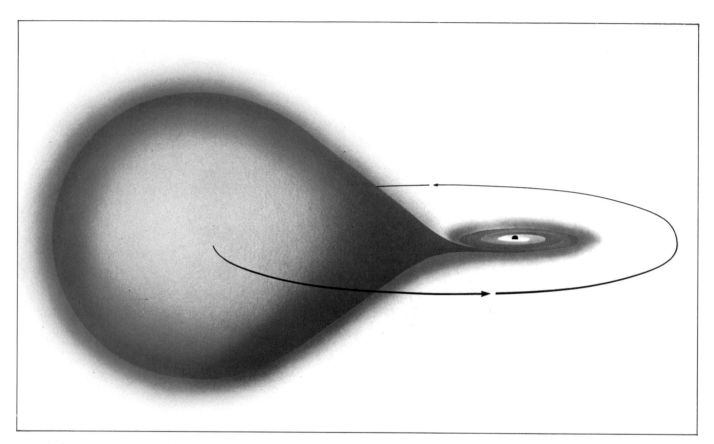

lord and master as faithfully as it orbits the bright Sun now.

Still thinking of the solar system, a comet, pulled in from the depths of space by the Sun's gravitation, usually has enough 'sideways' motion to miss the Sun's disc, and it is simply swung around by the close approach before heading back into the outer reaches of the solar system. And in exactly the same way, a spaceship approaching an invisible black hole in space is very unlikely to be heading straight at it (the Sun as a black hole would be only 6 kilometres [less than 4 miles] across, for example) and the space traveller would simply find his ship speeded up, swung round the invisible collapsed star, and sent back into space along a new path – leaving him perplexed and shaken up, but still in our Universe!

The invisible dance partner

But are there actually black holes in our Galaxy? The theory of star evolution certainly suggests it, for unless all heavy stars lose over 90% of their mass, their cores cannot stop as white dwarfs or neutron stars. Direct evidence is difficult to obtain, for black holes are very small, and – by definition – invisible. Nevertheless, astronomers are coming to believe that certain sources of X-rays in our Galaxy are related to black holes, and here we meet again one of the astronomer's most useful objects: the double stars.

Earlier on, we were introduced to double stars – two stars in orbit about one another – as a vital weighing machine for stars. Only in such a gravitationally-bound waltz is it possible to measure the masses of stars directly. But a star in a close pair can affect the way the other star evolves, and now that astronomers have fairly well sorted out the evolution of single stars, they are realizing that many of the most exciting events in our Galaxy are caused by the growing pains of one star being inflicted on a near neighbour.

Best known of these cataclysmic events is the nova explosion. The sudden appearance of a temporary 'new star' in the sky is a commonly recorded event in the written annals of many civilizations: the Chinese called them 'guest stars'; while in the West such an outburst takes its name from the Latin for 'new'. Every ten years or so, a nova becomes easily visible to the unaided eye, and many more faint ones are picked up by amateur sky watchers patrolling the sky with binoculars and small telescopes. During its outburst, a nova brightens up to a luminosity tens of thousands of times brighter than its normal state, and then gradually fades over the succeeding months. Statistics of nova outbursts suggest that such an explosion is simply one in a long series; a nova-prone star repeats its outburst at roughly 100,000-year intervals. Related variable stars, the dwarf novae, suffer lesser explosions more frequently.

In the past few years, observations and theory have neatly fitted together to explain these stars. In all cases, it seems, the original 'star' is, in fact, a very close pair, one of which is expanding into a giant, and the other has already gone past that stage and is a white dwarf. The expanding star has its problems, though. Its outermost gas layers on the side nearer the white dwarf are gradually getting closer to the dwarf, and further from the star's centre. There is a gravitational tug of war over this gas, and eventually the star reaches a size such that the white dwarf's gravitation pulls off its outermost layers; as the star continues to expand, the dwarf pulls off a fairly continuous stream of gas. Because the stars are continuously orbiting one another, this gas has a certain velocity sideways which prevents it from falling straight on to the dwarf; instead, it orbits the dwarf.

From close up, the ring of gas – the *accretion disc* – would look very much like Saturn's rings, but glowing with its own light. There is one very important difference, for where the stream of gas pouring off the giant star joins the gas already in the disc, the collision produces a very bright spot, which (in the dwarf-nova systems) can outshine the two stars themselves. Usually, then, the white dwarf seems to be surrounded by a faint gaseous disc, with an intense, white, flickering hot spot at one point in its outer edge. But every few weeks, the whole disc brightens up (for reasons which aren't completely understood), and a close-up observer would be dazzled by a brilliant ring around the white dwarf star; while, from our distance, we see a 'dwarf nova' outburst.

These gas accretion discs don't behave quite like Saturn's rings, which comprise many solid chunks moving in independent orbits. The gas atoms are continually colliding with one another, and as a result they are transferring angular momentum ('spin') outwards from the central parts of the disc. The matter in the disc is thus continuously, though slowly, spiralling inwards; and it eventually falls on to the white dwarf's surface. This pile-up of hydrogen-rich gas becomes degenerate – supported by electron pressure, like the white dwarf itself – but there is a drawback. Degenerate matter can have any type of nucleus to balance the charge of the supporting electrons except hydrogen nuclei, simple protons. In these high-pressure conditions protons combine to make helium, and they combine explosively. The accumulated hydrogen is blown off into space as an expanding shell, which temporarily far outshines the original star system, as a classical nova outburst. The system eventually settles down again, and hydrogen accumulates for another 100,000 years or so, until enough has piled up to fire another eruption.

What has this to do with black holes? The accretion disc in a binary system is the key link; but before looking at discs around black holes, let's investigate the intermediate case of neutron stars in a double system.

Imagine the previous scenario with the white dwarf replaced by a neutron star. Since the compact source is much smaller, the accretion disc extends further inwards, and here the gas is whirling around much faster, and it is much hotter; so hot, in fact, that it radiates X-rays strongly.

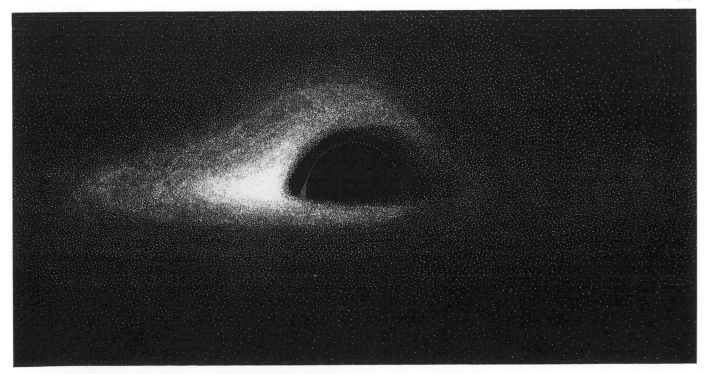

Over a hundred X-ray sources in our Galaxy have been detected by orbiting X-ray astronomy satellites, and the majority are of just this type (or of the closely related neutron star–supergiant combination, where the neutron star picks up gas shed by the latter as a stellar wind, rather than as a result of a gravitational tussle).

The neutron star's strong magnetic field causes added complications, however. Hot gas flowing down from the inner accretion disc is channelled to the magnetic poles to create X-ray hot spots near the neutron star's surface. As the star spins, these hot spots flash X-rays towards us (in much the same way as the radio waves from pulsars, although they are produced differently), and the X-rays come in pulses. The 'X-ray pulsars' discovered by X-ray satellites confirm the theorists' prediction that a neutron star is responsible for most X-ray sources. X-ray sources vary in many other ways too: some have short, repeated bursts; some 'flicker' continuously; some brighten and fade, rather like optical novae. X-ray astronomers are confident, however, that these can all be explained as variants of the original model of a neutron star in a close binary.

But there is always an exception. And the 'rogue' X-ray source is called Cyg X-1, the first X-ray source detected in the constellation Cygnus. At first this seemed to be a typical X-ray double-star system, although its position wasn't reliably known – as always happens in the birth of a new branch of astronomy. But in 1971, radio astronomers picked up this object as a radio source, and precisely located its position. It coincided with a supergiant blue star, catalogued as HDE 226868; and optical astronomers soon found that this star's spectral lines shifted backwards and forwards regularly, showing that its velocity was regularly changing. It was clearly orbiting an unseen companion star every $5\frac{1}{2}$ days. So far, so good: for this type of close orbit is just what astronomers expected for an X-ray source. But there was a surprise in store: when they calculated the mass of the unseen companion, it turned out to be at least five times heavier than the Sun.

Now there is a remorseless chain of argument, leading to just one conclusion. To get X-rays from an accretion disc, the central star must be small. White dwarfs are just small enough to be weak X-ray sources; but virtually all strong X-ray emitters are associated with neutron stars. Yet a neutron star – or a white dwarf, for that matter – cannot be as heavy as five Suns. The only object small enough and with this mass is a black hole.

It might seem contradictory to say that the Cyg X-1 X-ray source is a black hole: after all, no radiation can escape from within the event horizon of a collapsed star. In fact, though, it's only a confusion in language, for the X-rays come from the accretion disc of gas orbiting the hole. As we noted before, a black hole doesn't 'suck in' matter until it

comes right down to the event horizon. The gas in the disc is orbiting well outside this radius, and the accretion disc feels a very similar gravitational pull to that experienced by a disc orbiting a neutron star. Since a black hole is smaller than a neutron star of the same mass, the gas has further to spiral in, and so it gets hotter and emits radiation more freely; until, that is, it reaches the event horizon and disappears from sight.

And so the X-ray source Cyg X-1 is, strictly speaking, the accretion disc surrounding the compact companion of the supergiant star HDE 226868. But is this 'compact object' definitely a black hole? Many astronomers say: 'Yes, the large mass proves it; and besides, there is no sign of regular X-ray pulses, which we might expect if a neutron star were involved.' Other astronomers, not so happy with the idea of admitting the weird and wonderful black holes into our Universe, have tried to think of other explanations: perhaps the companion is actually two very close neutron stars; perhaps HDE 226868 is not a supergiant, but a fainter and nearer star mimicking a supergiant's optical spectrum. The majority of astronomers have been prepared to sit on the fence and see how future evidence fits in.

X-ray astronomers themselves are hunting for more X-ray sources like Cyg X-1, and the big High-Energy

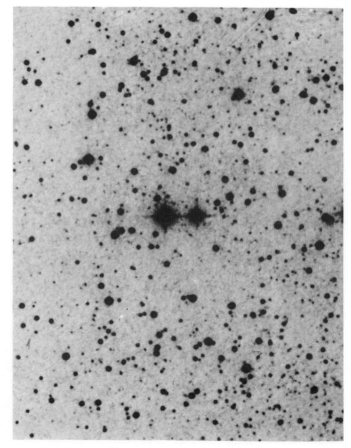

Astrophysical Observatory satellites promise to scour the sky effectively. These new satellites, the size of medium-sized vans, are ten times as big as their predecessors. Since they can collect far more X-rays, their detectors will pick up many more faint X-ray sources, and pin down their positions accurately enough for optical astronomers to look for companion stars. Already, another likely black-hole candidate has turned up in the constellation Circinus.

But collapsed stars are not the only possible black holes. Much larger and much smaller black holes may exist; and there is good evidence for a very large black hole in the centre of another galaxy. The colossal energy of the very distant and bright galaxies called quasars is probably related to black holes, too – but more of these in the next chapter. The gradually accumulating weight of evidence is forcing the majority of astronomers to come to terms with the fact that black holes probably do exist in our Universe. The problem is to find them.

Exploiting black holes

A black hole is invisible, by definition. Its only effects on the rest of the Universe are caused by its gravitation, its electric charge, and its spin. And in fact these are the only three things it's possible to know about a black hole, from outside it. If we squeezed together a thousand tons of water to make a black hole, there's no way we could distinguish it from a black hole made from a thousand tons of old cars. Even more surprisingly, a black hole made of antimatter is absolutely identical to one constructed of matter. The mathematics involved in black-hole physics is so difficult that it took several years for some of the best physicists in the world to prove this theorem – generally known as 'a black hole has no hair'.

Naturally occurring black holes probably have very little electric charge, for the matter collapsing into them will be electrically neutral. But judging by the rapid spin of neutron-star pulsars, collapsing stars are spinning extremely fast. Calculations show that the empty space around a spinning black hole is itself 'distorted'. If we watched a rocket falling straight in towards the centre of a rotating black hole, the distorted nature of space around it would start pulling the rocket in the direction of rotation, and the rocket would be forced into an infalling spiral path. Even better, suppose we had the technology to build a huge steel frame around the black hole, some distance away, and completely symmetrical so as not to be pulled in. The rotating space around the hole would make the frame turn, and we could thus produce a colossal amount of energy. One collapsed star could provide more than enough energy to supply our present level of consumption for longer than the age of the Universe – thousands of millions of times longer, in fact.

In another theorists' dream, we could send an unmanned rocket into the distorted space swirling around the black hole and, while it's there, eject a large section in a carefully controlled way. If this splitting up were done carefully it would be possible to make the ejected section shoot away from the distorted space as the rest of the rocket falls through the event horizon. 'Caught' in the right way, the speed of this fast-moving flotsam could be extracted as energy.

In both these ways, we could, theoretically, at least, extract energy from rotating black holes. In the second example, we could arrange to get out more mass-energy than the hole gains by swallowing the rocket. There is a limit, however. Extracting its energy of rotation naturally slows down the spinning black hole, and eventually it will come to rest, as a simple gravitational trap.

This leads us to another way of getting energy, not from the black hole itself, but from matter falling into it. Cyg X-1 may well be Nature's prototype of this kind of power station. As the gas around the black hole spirals inwards, it radiates X-rays; and this energy basically comes from the gravitational field of the hole. When a quantity of gas, or anything else, falls into a black hole, it can give out an amount of energy equivalent to almost half its own mass. One kilogramme of gas, for example, can be persuaded to radiate away 420 grammes as energy (almost half its weight), leaving 580 grammes to eventually fall through the event horizon. The black hole gets heavier by the latter amount, but we get a good return in energy for the investment in mass – in the Sun, one kilogramme of hydrogen fuses to helium liberating only seven grammes of energy.

How about producing a small black hole in the laboratory, and throwing in all our rubbish to be converted to energy? We could ensure that the hole has an electric charge, and hold it up against the downpull of Earth's gravity in an electric field (just like Millikan's early experiments with charged oil drops). Such a set-up could generate as much power as a large conventional power station on just 500 grammes (about 1 lb) of rubbish per year – it would solve our energy problems, but not our refuse disposal ones!

In fact there are insuperable difficulties in exploiting black holes on Earth. For a start, there's no way at present known for squeezing matter down to the small size and incredibly high density needed. We would have to squeeze a mass of a million tons down to the size of an atomic nucleus to make its surface escape velocity equal to the speed of light.

There's another, even more fundamental reason why such a tiny black-hole power station is not really feasible. For, to be completely contrary, it turns out that black holes are not really black. A million-ton black hole would be glowing at a temperature of a thousand million degrees, hotter than anything so far known in astronomy. And within a few years it would explode, in a cataclysm which would make a hydrogen bomb look like a damp firecracker.

Exploding black holes

How can a black hole not be black, and how can it explode, when its gravitational field is strong enough to keep even light within its event horizon? The original, simple idea of a black hole was overthrown in 1974, by incorporating the previously neglected ideas of the quantum theory. The resulting revolution has added strange new features to the already exotic black hole concept. And most surprising of all, perhaps, is the fact that this fusion of two most intricate physical theories – quantum theory and general relativity – was done entirely in the head, by the remarkable Cambridge theorist Stephen Hawking.

Although only in his mid-thirties, Hawking is already credited with having made the most important advances in our understanding of gravitation since Einstein himself. A muscular wasting disease has confined him to a wheelchair for several years, and prevents him from writing down his ideas; yet his capacity for original thought has, if anything, been sharpened by adversity, since he must keep all his ideas in his head and work them out by super-mental arithmetic. He's known in Cambridge as much for his sense of humour as for his bold and original theories about black holes.

Late in 1973, Hawking made his most important discovery. Einstein's equations for gravitation regard matter just as a mass which produces a gravitational pull. They ignore all the bizarre phenomena which take place on a very small scale: the wave-particle duality of matter; the Uncertainty Principle; the creation and destruction of matter and antimatter. When calculating astronomical problems, like the orbit of Mercury, such refinements are not needed; but Hawking was looking into the properties of very small black holes, where gravitation is intense and changes sharply over very small distances. And his quantum-theory approach seemed to show that the hole would emit elementary particles into space. At first, he admitted, 'I thought there must be some mistake.' But there wasn't; and his revolution burst upon the scientific world in March 1974, ensuring Hawking a leading place among those who have contributed to our understanding of the Universe.

The Uncertainty Principle, that strange law which tells us that we cannot know the instantaneous position and speed of any particle with complete accuracy, also has a ruling over energy. It is impossible to measure the precise energy possessed by any body or system, at an exactly known time. To put it another way, a small packet of energy could suddenly come into existence in empty space, but as long as it disappeared quickly enough, we could not detect it, and the law of conservation of energy would not be broken. (This is exactly the situation which allows the forces of nature to operate by the exchange of virtual particles, as

we found in Chapter 6.) This short-lived energy packet could turn into a particle-antiparticle pair (usually electron and positron), and normally the pair would almost instantly annihilate each other; and again, because it happens so quickly, we can't know of their brief existence.

But if a matter-antimatter pair springs into existence in the region of space just outside a black hole, one of the pair may fall into the hole, and so cannot cancel out the other in the usual way. The latter, bereft, particle travels out into space. From a distance, the black hole seems to be emitting individual electrons and positrons, which really come from the region around it. The positive energy of the escaping particle is balanced by the 'negative' energy which the black hole provides on absorbing the antiparticle. The 'free' particles are, in effect, carrying mass away from the black hole, and so the black hole must shrink in mass and size. Eventually, all the mass of the black hole will 'evaporate' in this unorthodox way. Black holes are not the final dead end of matter: although they effectively destroy the identity of anything which falls in, and nothing which has fallen in can get out again intact, they eventually dissociate themselves into a shower of the simplest possible particles. It seems poetic justice that the core of a dead, massive star, which started off as a mainly proton and electron (hydrogen) gas, should end up – after all its vicissitudes as nuclei with the complexity of iron, and the ignominy of total gravitational collapse – as an outburst of new fundamental particles.

Putting figures into the calculations does show, however, that the previous calculations of collapsed star cores are for practical purposes all right. Hawking's particle production is most effective when the gravitational field changes markedly over only a small distance, and this means that it becomes less and less important as we go to larger and heavier black holes. The black hole in Cyg X-1 would take a time far longer than the present age of the Universe to disappear: in fact at present it is pulling in immensely more gas from its companion than it is losing from 'evaporation' of fundamental particles.

But the situation is dramatically different for the smaller holes that Hawking was working on at the time. In the inferno of the 'Big Bang' in which the Universe began, some 15,000 million years ago, it's quite possible that a whole range of small black holes was produced. The smallest ones will already have evaporated, but those with a certain mass will be disappearing now. Since 'evaporation' makes the black holes less massive, and therefore more prolific in their particle output, the outward stream increases with time, and the final disappearance takes place in a fraction of a second with an immense burst of energy. The black hole literally explodes. Astronomers are now searching for bursts of gamma-rays and radio waves which

might be coming from such black-hole explosions: if they find them, it will be a tremendous triumph for Hawking's insight into the exotic workings of the Universe.

The energies of the particles streaming out from a black hole mimic the distribution that we get from a hot shining 'black body' as radiation. So it's possible to say that a black hole is, effectively, at a particular temperature. A black hole with the Sun's mass – some 6 kilometres (3·7 miles) across – is effectively only a ten-millionth of a degree above the absolute zero of temperature. A Moon-mass hole – 0·25 millimetres (0·0098 inch) across – would be comfortably warm, while the thousand-million-ton black holes that may be exploding now are at some 120 thousand million degrees. So black holes can't be regarded as black. Although they swallow all light coming into them, there is a thin region just outside the event horizon which is continuously 'shining', and draining the hole of its accumulated energy.

What goes on inside a black hole? We'll ignore the *spinning* black holes, for theory indicates that these may whisk us into another Universe. Although theorists can quite happily describe 'our Universe' and 'other universes' in equations, and say that they can be connected by spinning black holes, it's actually not at all certain that a spaceship venturing into such a hole would really find itself emerging in another universe full of different galaxies. If we flew straight at a non-spinning hole, once inside we could see what happens to the core of our massive collapsing star. No known force could stop it; and Stephen Hawking and his mentor, mathematician Roger Penrose, proved in 1969 that the inevitable result is a 'singularity'. All the matter collapses to literally zero volume. A singularity is a mathematical point in space, where a mass is compressed to infinite density in a region exactly zero in size. (This theorem, incidentally, also proves that once our spaceship is within the black hole, it must end up in the singularity too, infinitely squashed.)

Singularities are even more of an embarrassment to physicists than are the black holes surrounding them. Some theorists have tried to prove that singularities are always surrounded by a black hole, and so are hidden from our view. But this cosmic censorship won't do: naked singularities probably can occur, and theorists must try to explain them.

There's one possible way out. As John Wheeler of Princeton University is fond of pointing out, these final singularities arise from a gravitational theory which ignores the quantum effects of matter; yet when we compress matter in this way, to very small sizes and very high energies, the quantum effects must become extremely important. Like the other forces, gravitation itself must be conveyed by 'force particles', the gravitons; but these are just 'smeared out' by the blanket approach of general relativity, which treats gravitation as a distortion of space and time. But it's extremely difficult to combine the two different approaches to the same force. There is, too, the hope of combining gravitation with the other forces, as different aspects of one all-embracing force.

Will the embarrassing prediction of singularities in the Universe disappear, when, as Wheeler puts it, 'the fiery marriage of general relativity with quantum physics has been consummated'? Does the centre of a dead star actually end up as a zero-sized point in space, or will the new physics predict some strange small object instead?

The field of black-hole studies is perhaps the most intri-cate mixture of astronomy and physics. Astronomers are searching for large black holes by the effects of their gravitational attraction, and small black holes by their explosions. It's important to know if our Universe does actually contain black holes, and the gradually increasing evidence for the larger holes is of paramount importance in understanding gravitation.

Physics, meanwhile, can guide astronomers in their quest; and it can provide the answers to some of those questions which can never be answered experimentally, such as revealing what actually goes on in the forever unsee-able interior of a black hole.

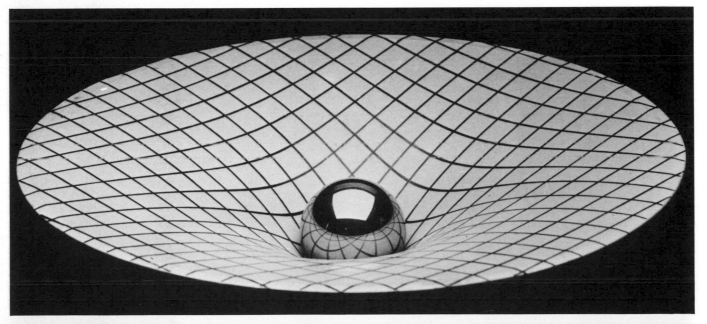

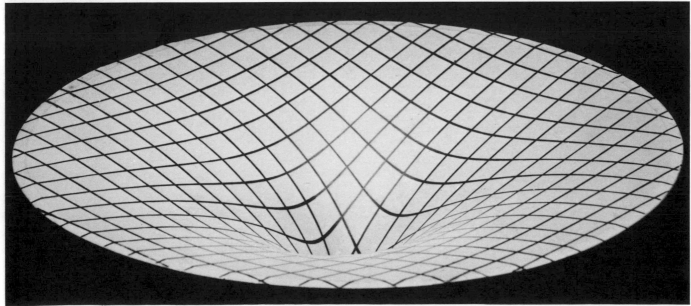

The Andromeda Galaxy is the nearest large galaxy to ours, 'only' 2¼ million light years away. This diffuse spiral is composed of some 200,000,000,000 stars, too faint to be seen individually. Appearing to the unaided eye as a faint misty patch, the Andromeda Galaxy is the most distant object visible without a telescope. Copyright by the California Institute of Technology and the Carnegie Institution of Washington. Reproduced by permission from the Hale Observatories.

10:Geography of the Universe

We live in a huge, slowly rotating Catherine wheel of stars: the Milky Way Galaxy. Our Sun is a fairly average member of our Galaxy's family of some 100,000 million stars; our nearest neighbours shine out as twinkling points in the night sky, while the distant cohorts, too faint to be seen individually, pile up in perspective as the glowing band of the Milky Way. Astronomers have analyzed the Milky Way star family in detail, and the story of a star's life and death has been pieced together by studying the different kinds of stars in our Galaxy. All the complexities of gas clouds collapsing into stars, and stars exploding back into gas and dust – and the exotic neutron stars and black holes – are deduced by astronomers looking at the contents of our Galaxy.

But our Galaxy is only one of millions of galaxies in the Universe. Seen from the southern hemisphere, our two nearest neighbours, the Magellanic Clouds, shine out as fuzzy cloud-like patches behind the veil of nearby stars in our own Galaxy. And northern observers can pick out the fainter oval of the Andromeda Galaxy on clear autumn evenings. This galaxy is the furthest object we can see with the unaided eye, two million light years away from us. It's a huge galaxy, made up of twice as many stars as the Milky Way Galaxy, but reduced by its distance to a mere glow in the sky.

Telescopes, of course, reveal many more galaxies, especially on long-exposure photographs which bring out their faint outer regions. Galaxies come in all sizes; and the smallest ones in all shapes, too. But larger galaxies are either spirals, or else the oval, fuzzy balls of elliptical galaxies. Spirals are galaxies with character: after a little practice an astronomer can recognize individual galaxies by the details of the structure in their winding arms, with much the same sense of recognition as when meeting old acquaintances in the street. But the plain-Jane ellipticals are much more anonymous, and few astronomers could unerringly identify a particular elliptical galaxy from its picture alone.

The differences between the galaxy types – the small irregular ones, the large spirals, and the ellipticals, which come in all sizes – depend on how they were originally formed, as we'll see later. But the similarities are perhaps more surprising than the differences. For a start, the spectroscope shows that galaxy spectra are crossed by dark lines, corresponding to the 'fingerprints' of elements which occur in stars of our Galaxy. Indeed, computer studies have shown that it's quite easy to mimic a galaxy spectrum simply by combining the light from types of stars which are common in the Milky Way Galaxy. So the other galaxies are made of stars very like the stars in our own – and for the nearest galaxies, astronomers have checked this directly. Individual bright stars show up on photographs of the

Magellanic Clouds and the Andromeda Galaxy (and other nearby galaxies); and they have the colours and brightness that we would expect.

Other spiral galaxies, and the irregulars, contain gas and dust between the stars, and again the spectroscope shows it to be very like the gas and dust in our Galaxy. So the Milky Way is a very average galaxy, nothing special in the cosmic scene. This is a very useful piece of information to astronomers, because the stars and gas in our Galaxy are pretty well understood, and there seems no reason why we shouldn't transfer our knowledge wholesale to apply to the constituents of other galaxies. And there's feedback the other way. By looking at all the types of galaxy in the Universe, and finding their similarities and differences, we can see how the Milky Way fits in, and learn something about why it is the type of galaxy that it is – just as we can understand more about the Earth by comparing it with the other planets of the Sun's family, and can find out about the Sun's past and future by studying the other stars in our Galaxy.

A galaxy like ours contains stars of all ages, as does any spiral or irregular galaxy. Young stars are being born right now in huge gas and dust clouds, such as that near the Orion Nebula. The Sun is some 5,000 million years old, and many stars are older still. But there's a limit at an age of about 15,000 million years. We could detect stars – in practice, star clusters – older than this; but there just aren't any such very ancient star clusters in our Galaxy; astronomers conclude that our Galaxy must have come into being some 15,000 million years ago. And here's another remarkable similarity, for virtually all other galaxies seem to be roughly the same age. It isn't a measurement that astronomers can make at all accurately, but they are confident that there are no extremely old galaxies. Recent research has highlighted some galaxies which may have formed more recently, but such galaxies are rare and fall into the insignificant 'dwarf' category. The major galaxies all came into existence at much the same time.

All the galaxies are identical when it comes down to the fundamental physicists' level. They are all made of the same fundamental particles, acted upon by the same forces. As a result, they have made up similar gas and dust clouds, and similar stars. To the physicist, the contrast between the beautiful arms of a spiral galaxy and the plainness of an elliptical is a very minor difference, which can be explained as a detail of galaxy formation. The very similarity of the basic ingredients of galaxies, and their nearly identical ages, indicates to astronomers that they formed from an originally continuous cloud of gas which filled the Universe. Some 15,000 million years ago this 'cloud' (perhaps 'medium' would be a better word, for it filled the whole Universe) started breaking up into clumps, each collapsing inwards under its own gravitational attraction to make a galaxy, and leaving virtually empty space in between.

The birth of the Milky Way

Our Galaxy began as a huge cloud of hydrogen and helium, some 300,000 light years across – just one of the clumps which gravitation was coalescing out of the general gas filling the Universe. As it shrank in size under the force of its own inpulling gravitation, this 'protogalaxy' (forerunner of a galaxy) became denser and denser, until the gas began to fragment into a myriad of smaller lumps which collapsed into the first stars of the Milky Way Galaxy.

These were rather odd stars, for they consisted only of hydrogen (three-quarters by weight) and helium (one-quarter). Later stars, like the Sun, have a small admixture of heavier elements – carbon, silicon and iron, for example – which have been made in, and scattered by, previous supernovae. These elements also make up the grains of dust in interstellar space; and when present-day stars form in dense gas and dust clouds, these dust grains play an

important role in cooling down the central gas and letting it collapse into stars. But at the beginning of our Galaxy, the only elements in existence were the two lightest, hydrogen and helium; and the first stars cannot have been born in the seclusion of an all-enveloping dust curtain. Indeed, astronomers are not yet certain exactly how these first stars did arise; but anyway, once they had formed, nuclear reactions at their hearts began to convert the basic 'fuels' into heavier elements, and the more massive stars soon exploded as supernovae, spreading the heavier elements through the Galaxy, and starting to build up the mixture of elements that we see in space today.

The protogalaxy must have started off fairly spherical, a huge collapsing globe of gas. But it was rotating slowly. Gravitation, trying to pull all the gas in towards the centre, had to compete with the physical law which states that the quantity of spin (angular momentum) must always stay the same. As the gradually turning cloud shrank, it spun faster; and instead of all the gas falling to the centre, it ended up as a rotating disc. This was the only compromise that gravitation and angular momentum could amicably reach. The total effect of the matter in the disc is a complicated distribution of gravitational field, but the overall gravitational attraction essentially pulls every part of the disc towards the centre; and this inwards force is counterbalanced by the rotation of the disc. It's rather like the revolution of planets about the Sun, but here there isn't just one very heavy body at the centre: every star and gas cloud in the disc feels the gravitational pull of the others, and these average out to an inward pull. Our Galaxy's spiral arms are part of this disc, and the Sun is one of the stars in it; but before we look at this part of the Galaxy, there is an older region to survey.

In the outer reaches of the original, spherical protogalaxy, stars began to form before the gas collapsed to a disc. And there's a very important difference between the way stars and gas move in a galaxy. The atoms in a gas are all moving at high speed, and constantly colliding, so speeds and motions very quickly average out. That's why the infalling protogalaxy flattened itself out into a smoothly rotating disc: any gas atoms going the 'wrong way' at first would soon be bumped into moving the same way as the majority. And this collapse took place very quickly on the astronomical time scale – in only about a hundred million years.

But the gas which turned into stars early on suffered a different fate. Stars orbit the Galaxy without ever closely approaching one another. Think of how small the Sun is, compared to the distances between the stars: if we scaled the stars down to human size, it would mean placing people an average of 10,000 kilometres (6,000 miles) apart, and the chance of two people accidentally meeting during a lifetime would be correspondingly minute. So once a star is born, only the average gravitational pull of the whole Galaxy affects it. The stars which formed when the Galaxy was a huge ball still pursue orbits which take them right up and down, above and below the disc region of our Galaxy, tramping their way through a large spherical region which astronomers call the Galaxy's *halo*.

These halo stars can be picked out in several ways. They always contain less of the heavy elements at the surface than do stars of the disc, simply because they were born earlier. And their long orbits about the Galaxy can be arranged in any old way in space, only occasionally taking them through the disc. As the Sun whirls around the galactic centre with the other disc stars, we are steaming past any halo stars which happen to be passing through the disc near us. To us it seems that these slow-moving stars are moving rapidly in the other direction; and they've gained the entirely misleading name of 'high-velocity stars', when in fact it is the Sun which has the high velocity. Before astronomers had properly disentangled the way our Galaxy is constructed, it seemed almost astrological that there was a link between a star's chemical composition and its apparent speed through space.

About a tenth of the halo stars are not pursuing individual orbits, but are bound up in giant clusters of up to a million stars. These large globular clusters are, naturally, the most obvious things in the halo. Despite their distances from the disc, typically some tens of thousands of light years, we can see some of these outer bastions of our Galaxy even without a telescope. From the Southern Hemisphere, two are visible as 'fuzzy stars'; they have indeed been named as stars, Omega Centauri and 47 Tucanae. In the northern constellation of Hercules, there's a fainter one, which can be seen on a really clear night, and bears the more cryptic name M13.

This indicates that it is the thirteenth object in the catalogue compiled by Charles Messier, an eighteenth-century French comet hunter, nicknamed by Louis XV 'the ferret of comets'. Messier became increasingly annoyed by the number of nebulous objects in the sky which could be taken for faint comets at first sight, and the upshot was his famous catalogue of 103 'nebulae'. Some of these we now know to be genuine gaseous nebulae, like the Orion Nebula (M42), some are galaxies (the Andromeda galaxy is M31), and some are star clusters like M13 or the Pleiades (M45). Messier's numbers are still a useful way of identifying the brighter nebulae, clusters and galaxies, even though his catalogue has long been superseded by much more comprehensive ones for specialist use. It's ironic that Messier is now remembered not for his many comet discoveries (thirteen in all), but for his list of celestial nuisances!

The globular clusters are the oldest objects you can see.

The stars in them have been shining steadily for some 15,000 million years, three times longer than the Sun's existence, and almost since the beginning of the Universe itself. (Our Galaxy formed soon after the Universe's birth, as we'll see in the next chapter.) Astronomers are particularly glad that there are globular clusters to date the Galaxy's formation by, because it's not easy to measure the ages of individual halo stars. With clusters, though, it's relatively easy – with the accent on 'relatively'.

When the cluster first forms out of a gas cloud, stars of all different masses appear, and begin to fuse hydrogen to helium in their cores. But stars have limited lifetimes, and as we saw in Chapter 8, the spendthrift heavyweight stars get through their 'fuel' at a prodigious rate, and blow themselves apart as supernovae within a few tens of millions of years. Clearly, the globular clusters won't contain any of these stars now. A star with the Sun's mass lasts around 10,000 million years, and a Sun-like star in a globular cluster aged 15,000 million years will already have gone through its entire evolution and will now be an insignificant white-dwarf ember. Stars slightly less heavy will be going through their enormous red-giant stages, and only when we go down to very lightweight stars do we find any that are still fusing hydrogen at their cores. These low-mass stars are

A wide-field view of the centre of our Galaxy, in the constellation Sagittarius. Nearby dust lanes in the disc make a dark obscuring band across this bulge of old stars – and hide the very small

central nucleus detected by radio and infrared astronomers. The streak is an Earth satellite which crossed the field during the exposure.

consuming hydrogen so slowly that their stocks have lasted virtually the age of the Universe.

Theory tells us how long the hydrogen supply of a hydrogen-fusion star (a 'main-sequence' star, in astronomers' jargon) will last. So we look at a cluster, and find the heaviest hydrogen-fusion star there, by plotting a Hertzsprung-Russell diagram of each star's luminosity against its temperature, and following the main sequence up from the lower right corner until it abruptly ends. Stars at the end of the main sequence are just finishing their hydrogen fusion; their cores are about to collapse, with the swelling of the whole star to a red giant. From the theory, we know the time it has taken a star at this position on the main sequence (which corresponds to a particular mass) to fuse all its central hydrogen supply; and this time is therefore the age of the cluster.

As mentioned earlier, the oldest clusters turn out to be around 15,000 million years old. This must have been the time when our Galaxy separated out from the general gas

filling the Universe, and the globular clusters and other halo stars were soon left stranded. Pursuing orbits which carry them eternally throughout the huge original domain of our protogalaxy, these first stars were left behind as the other nine-tenths of the gas sank down to form the Galaxy's disc. Until a few years ago, astronomers viewed the halo as a minor part of the Galaxy, interesting only because its stars are so old. But there's now a growing feeling among some astronomers that we may have underestimated our Galaxy's halo: instead of being only one-tenth the mass of the disc, it may be ten times heavier.

The first indication came from theorists who were studying the spiral arms which lie in the disc of the Milky Way. They made the embarrassing discovery that their equations just didn't produce long-lasting spiral arms unless the Galaxy's halo had a strong gravitational effect. This meant that the huge halo region must contain at least as much matter as the disc; and later studies of globular clusters and small galaxies near ours showed that the Milky

Way Galaxy seems to have a much stronger gravitational pull than it should. The only place where this extra mass can be 'hiding' is in the extensive halo, above, below and around the disc. Our picture of the Milky Way could thus change from being an all-important disc of gas and stars, surrounded by a few older stars, to a huge spherical region, making up most of the Galaxy's mass, which contains a relatively unimportant disc, prominent only because its constituents shine brightly.

Many astronomers are wary of this new 'massive halo' picture, though, for the simple reason that we can't actually see all the supposed extra matter. For every star in the halo, there would have to be a hundred times as much matter, which just isn't visible. What is the extra matter? It's certainly not there as ordinary stars or gas, which astronomers could detect with optical, radio or X-ray telescopes; nor as dust grains, which would block our view of the Universe beyond. But there are ways of putting matter in the halo in such a way that astronomers couldn't detect it. One obvious possibility is a vast number of black holes, by definition invisible; or, alternatively, a plethora of planet-sized objects, moving independently rather than orbiting a star. Bodies fewer and larger than dust grains won't block off a significant amount of light from beyond the Galaxy; and, as long as they are lighter than fifty Jupiters, they won't shine as stars. If the astronomers who believe in the massive halo are right, our Galaxy is not predominantly made of stars and gas clouds. These obvious constituents would in reality be only a minor component, and the Galaxy's bulk would be essentially unseen matter: black holes, planets, icebergs. It's a difficult controversy to settle, since astronomers can't just go out and make simple observations to decide the issue: the theorists have led trumps by declaring that the extra matter is invisible. Yet it's the most important question now being asked about galaxies in general, and our own Galaxy in particular.

The spiral arms of our Galaxy are revealed by the 21 cm radio emission from neutral hydrogen gas. This 'photograph' of the Galaxy in plan is centered on the Galaxy's centre; the Sun's

position is shown by the small circled dot near the top. (The blank portion at the bottom is due to the method used in making the map). Courtesy Gart Westerhout.

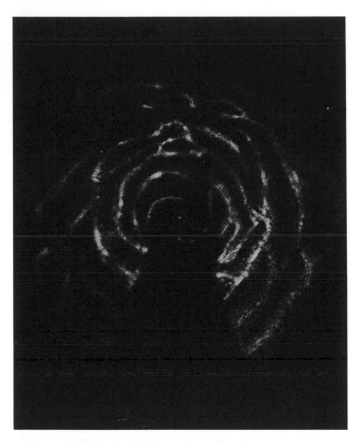

Let's leave behind the still-enigmatic halo now, and follow the fate of the rest of the protogalaxy's gas which settled down to form the rotating disc. Here stars were born; but not all the gas turned to stars. To this day, the disc is one-tenth gas, primordial gas mixed up with ejected debris from the supernova explosions of dying stars. Supernovae fling heavy elements into the gas; so the next stars to form have more of these elements, and they are surrounded by dust grains which can condense into planets.

The most obvious feature of our Galaxy's disc, as seen from outside, is that it doesn't look like a 'disc' at all. It would appear as a pair of spiral arms winding out from the Galaxy's centre, just as we see on photographs of other spiral galaxies. This is something of an illusion, for the spiral arms are strung with bright stars and glowing nebulae, which give them greater visibility. The fainter stars are spread more uniformly, between the arms as well as in them.

The spiral pattern seen in many galaxies was a beautiful enigma for over a century after the third Earl of Rosse first saw winding spiral arms in some nearby galaxies with his monster of a telescope. The problem is that the stars in a galaxy's disc don't all move around its centre at the same rate. If the spiral arms were always made of the same stars, the stars of one near the galaxy's centre would be circling

faster than stars further out in the same arm, and so the arms would gradually wind up, becoming more and more tightly coiled. This just doesn't tie in with the open arms we see in most spirals; and astronomers have eventually come up with an alternative theory, which seems to fit the observations well.

Suppose we start with a disc galaxy, where the stars (leaving aside the gas for the moment) are spread out through the disc, but are slightly concentrated into a two-armed spiral pattern. As the stars move around their orbits, those which start off in the arm move out of it into the general inter-arm region of the disc, while stars originally between the arms will orbit into an arm. So the stars which make up an arm are constantly changing; the spiral arm is just a pattern, showing up where stars are most strongly concentrated. You might expect the spiral pattern to become quickly smeared out as a result of the stars' motions; but here we meet the real beauty of the theory. Because originally the stars were more concentrated in the spiral arms, the arms have effectively a stronger gravitational pull than any other part of the disc. A star orbiting towards an arm is speeded up as it approaches the pattern, and slowed down as it leaves the far side. So, unlike a planet pursuing a circular orbit about the Sun, a star doesn't circle the galaxy at a constant rate. It spends longer than it should in the region of a spiral arm. But while the star is in the arm region, its gravitational pull is contributing to the overall pull of the arm: it helps to alter other stars' speeds so that they, too, will spend longer in the arm region.

The result is a self-perpetuating pattern. This assembly of two spiral arms actually rotates slowly, but at a different speed from that of the stars; so it always preserves its beautifully open form, without winding up. Look at a spiral arm now, and again in 50 million years' time, and the stars making it up will have changed; and the spiral will have turned slightly. But the overall pattern of two winding spiral arms has been handed down as the stars move through them: the gravitational pattern which guides the stars is passed on by the gravitation of these stars themselves – rather as the insubstantial myths of the ancients were passed unchanged down successive generations, to guide our ancestors' lives.

The spiral pattern has a more dramatic effect on the gas in the disc. The shock of a sudden speed change at the arm clumps the gas together in huge clouds strung along the inside of the arms, and the dust grains show up these clouds as a dark lining to the bright arms. The sudden squeeze can force the clouds to collapse further into stars, and photographs of other galaxies show the arms bejewelled with bright nebulae where stars have just formed. Extremely massive hydrogen-fusion stars also light up the arms, for these very bright stars live for such a short time before

165

exploding as supernovae that they don't even have time to orbit out of the arm where they are born. Spiral arms are exciting places, full of star birth, star death and bright and exotic stars.

Spirals of the Milky Way

Since we live in the thick of the Milky Way's disc, some 30,000 light years from its centre, it's not easy to see our Galaxy's spiral structure. In this respect, astronomers know less about our own Galaxy than about many others. One approach is to look for the kinds of object which live only in spiral arms – large nebulae and bright massive stars. In this way, astronomers have worked out that the Sun lies roughly half-way between the turns of our Galaxy's spiral arms. Some 7,000 light years further out lies the Perseus Arm, while the same distance towards the Galaxy's centre is the Sagittarius Arm. Near the Sun we have the bright young stars and nebulae so prominent in the constellation Orion, although astronomers now think that this 'Local Arm' is only a spur of the Perseus Arm. When the Sun orbits through a major arm, the entire sky must be as brilliant as the constellation Orion.

Radio astronomers have pushed out further, to lay bare the entire spiral structure of our Galaxy. Conveniently enough, the most common atom in space, hydrogen, emits radiation in the radio part of the spectrum. The hydrogen atom is simply an electron orbiting a proton, and both these particles spin on their axes. Quantum theory insists that their spin axes are lined up; so they can spin either in the same direction, or in opposite directions – but not in any other plane. Flipping from 'same-spin' to 'opposite-spin', the electron shoots out a quantum of energy: a photon with a wavelength of 21·106 centimetres. Although each hydrogen atom does this only rarely – once every 11 million years on average – there's enough hydrogen around to broadcast a measurable signal.

And because this is a spectral line whose wavelength is extremely accurately known, astronomers can measure the speed of any interstellar hydrogen cloud towards or away from us by looking for a Doppler shift in the wavelength it radiates. The Galaxy is rotating fast, by ordinary standards (the Sun orbits the galactic centre at 250 kilometres [155 miles] per second) and the rotational speed needed to balance the Galaxy's gravitational pull depends on a cloud or star's distance from the centre; so almost all the hydrogen is moving relative to us.

By tuning in carefully to the 21-centimetre broadcasts from hydrogen clouds, astronomers have come up with two important results. First, they can 'weigh' the Galaxy. Every star and gas cloud has its speed determined by the gravitational attraction of all the other stars and gas closer to the Galaxy's centre. Putting this in figures, astronomers

To an X-ray detecting telescope, the galaxies in a cluster are invisible. The X-rays arise in a cloud of very hot gas lying between the galaxies and condensed towards the cluster centre. This colour coded 'contour map' shows the X-rays emitted by gas in the Perseus cluster of galaxies. Courtesy John Zarnecki.

can find how the Galaxy's mass is distributed. Within the Sun's orbit, the total mass is about 100,000 million Suns; since the Sun is about two-thirds the way out to the edge of the disc, the total disc mass is probably some 50% larger. (Unfortunately, this method doesn't tell us anything about the massive halo which may or may not surround the disc.)

Turning these results around, once we know how the disc is rotating, astronomers can use the apparent speed of a gas cloud to calculate its distance. And so the dense gas clouds of the spiral arms can be plotted out on a chart to show the grand spiral design of our Milky Way Galaxy. Optical astronomers cannot see these distant reaches of the arms, because the dust grains in interstellar space block off their light; and so it was a significant step early in radio astronomy's history when the hydrogen-arm maps showed how much we could learn from radiations other than light.

Brothers in space

That's a quick overview of the Milky Way Galaxy, our home system, and a pretty typical spiral. Our nearest large neighbour, the Andromeda Galaxy, is almost a twin of ours, although about half as large again – it's one of the biggest spirals known. Basically all spirals are much like ours, with

dim haloes and bright discs: to realize the rich variety of beautiful forms which these galaxies can take, it's best just to look at the photographs.

Astronomers classify spirals according to the openness of their arms, and the size of the 'bulge' of old stars at the centre of the disc. It's most unlikely, incidentally, that galaxies change from one kind to another as they age. Galaxy shapes are probably fixed by the way the original protogalaxy collapses. There's one distinct type of spiral galaxy which is slightly different, and that's the class of *barred* spirals, where the central bulge of old stars forms a short bar. The stars here are in orbits which perpetuate the bar shape, rather as spiral arms are maintained. By the way, we can't rule out the possibility that our own Galaxy is a barred spiral; some astronomers believe that certain perplexing gas motions near the galactic centre are caused by the gravitational pull of a rotating bar.

The smallest spiral galaxies are about one-third as large as ours. There are smaller galaxies with roughly the same admixture of stars and gas, but these don't take on a spiral form. Remaining undefined in shape, they have been called irregular galaxies. The Small Magellanic Cloud, visible from southern latitudes, is one such small irregular galaxy. Its companion, the Large Magellanic Cloud, is a fascinating borderline case. It's almost big enough to be a spiral, and you can certainly convince yourself that it has a central bar and one faint spiral arm – some astronomers hedge their bets by classifying it as a one-armed 'Magellanic-type spiral'.

Both spirals and irregulars contain a lot of gas, forming stars right now. But there's another class of galaxy in the Universe, a type with no gas, and no star formation. As these galaxies first condensed, all the gas in the protogalaxy must have turned to stars at once, leaving none for future star generations – or else their residual gas has somehow been stripped away. These stars have orbited endlessly for the last 15,000 million years in a roughly spherical volume of space, making up a featureless, fuzzy-looking elliptical galaxy. Elliptical galaxies have no character, no individuality; until recent years astronomers saw them as merely faded relics of a long-lost grandeur. But new discoveries have revealed some ellipticals in a fresh light, as we shall see later.

First, though, let's look at the grand design of galaxies in space. They are not just scattered at random, but are grouped gregariously into clusters of all sizes. The Milky Way, along with the Andromeda Galaxy, a smaller galaxy in the constellation Triangulum, the Magellanic Clouds, and a couple of dozen dwarf galaxies make up our Local Group of galaxies. It's a small one, on the cosmic scale, but it's probably a fairly typical corner of the Universe. Exciting discoveries have been made even this close at hand on the extragalactic scale: radio astronomers have found a streamer

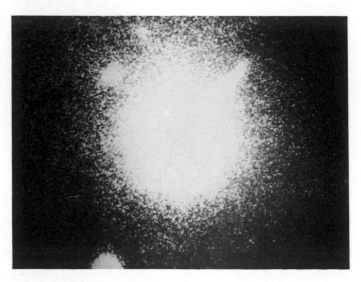

of hydrogen gas linking the Magellanic Clouds to our Galaxy, and trailing for 200,000 light years through the almost empty intergalactic space; the dwarf galaxies, so faint that they would be virtually invisible outside the Local Group, and yet probably the most common type in the Universe; and a giant elliptical galaxy, Maffei 1, which may be the largest member of our group, but was undetected until 1968 because it happens to lie behind obscuring dust in our Galaxy.

Other clusters of galaxies are larger, with up to three thousand members – galaxies all held together in a huge swarm by their mutual gravitational attraction. At the centre of the biggest clusters lurk the supergiant elliptical galaxies, the most massive galaxies in the Universe. Many astronomers believe that they have grown to this size – a hundred times heavier than the Milky Way – by cannibalism. Starting off as ordinary large ellipticals, they have swallowed up galaxies passing through the cluster centre, adding these stars to their own populations, and so swelling to enormous dimensions.

Lying between the galaxies in these rich clusters is a vast amount of hot gas – and 'hot' here means an unimaginable 100,000,000°.C Such gas emits X-rays copiously, though it produces little visible light, and X-ray astronomy satellites have picked up some fifty of these extremely distant gas clouds trapped by the gravitation of clusters of galaxies. This trapped gas probably explains why the galaxies in rich clusters are almost all ellipticals – a type rare in the Universe as a whole. Any galaxy moving through the cluster gas will be stripped of its own gas, to leave a galaxy of only old stars. In clusters, astronomers have spotted a sub-type of elliptical: lens-shaped galaxies of old stars, which seem to be the remains of ordinary spiral galaxies with all their gas stripped off in this way, leaving a disc of old stars alone. An ordinary spiral like our Milky Way would stand no chance in a rich cluster: if it escaped the predations of the supergiant cannibal galaxy at the centre, it would still be shorn of its beautiful spiral arms by the cluster gas.

There's still one big mystery about clusters of galaxies: the missing mass. Astronomers can calculate the masses of elliptical galaxies from the shapes of their spectral lines, widened by the Doppler shifts of all the stars in the galaxy; or else from pairs of ellipticals in orbit about each other. So it's quite easy to add up the masses of all the galaxies in a cluster, and see how strong their gravitational attraction should be between one another. But the galaxies in a cluster are moving too fast: the total gravitational pull just isn't enough to hold them together. Astronomers have concluded that there must be something else in the cluster to provide extra gravitational pull – and that this invisible matter must provide some five to ten times the mass which resides in the galaxies themselves.

The galaxies are distributed in such a way that the missing mass must be spread out through the cluster. Although the hot gas is spread out like this, the X-ray measurements show there just isn't enough of it; the missing mass seems to be in the form of a huge number of invisible bodies. If you think that the fact that we see only one-tenth of the matter in clusters seems remarkably similar to the idea that most of our Galaxy's mass may consist of an invisible halo, you are in good company. Some of today's more daring astronomers are indeed speculating that most of the matter in the Universe may be invisible – and that means to radio astronomers, X-ray astronomers and the rest, as well as to ordinary telescopes. The question is far from being settled yet, though; there may, in fact, be ways to solve the missing-mass problem without such a bold step; and astronomers are certainly not laggard in their attempts to understand just what is going on in these huge clusters, the largest congregations of matter in the Universe.

Quasars

In 1963, astronomers were rudely awakened from centuries-long contemplation of the heavens as a serene realm of harmony, with planets smoothly circling stars, stars genteelly orbiting in galaxies, and galaxies as peaceful islands in space. The young science of radio astronomy joined hands with optical astronomy, and the result was a shocking new discovery: the quasars.

Radio astronomers were at an early stage in their exploration of the radio waves from the sky. Their work consisted largely of cataloguing individual radio sources, and measuring their positions accurately so that optical astronomers could see what lay at that point in the sky. Already, sources in our Galaxy were known to coincide with hot nebulae (like the Orion Nebula) and with the expanding gas shells from old supernovae, the most famous being the Crab

Nebula. Some galaxies, too, seemed to be radio sources, and the second-brightest radio source in the sky – designated Cygnus A – was associated with a faint, distant galaxy. Since this source 'outshone', in terms of radio waves, thousands of nearer galaxies, something extraordinary must be going on there. Yet the astronomical world was not prepared for violence on the scale of the next discovery.

The quasar story started innocently enough, when accurate radio positions pinpointed three small radio sources as coinciding with starlike objects on photographic plates of the sky. Astronomers assumed that these points of light were indeed stars in our Galaxy, but stars of some peculiar kind which were powerful radio transmitters. When they took spectra of these stars, they turned out to be peculiar indeed. Instead of a light band of all colours, crossed by dark lines, these 'stars' had bright lines – and the wavelengths of the lines were even more surprising. As we saw in Chapter 7, each element emits and absorbs light of just a few specific wavelengths, which are altered only slightly by the Doppler effect when they arise in a moving star in our own Galaxy. But the lines of these starlike objects corresponded to no known element; and even a radio-emitting star shouldn't be that peculiar!

The key to the puzzle was provided by another radio source, catalogued number 273 in the third Cambridge catalogue of radio sources – 3C 273, for short. The Moon passed in front of this source in 1962; radio astronomers timed the exact moment when its radio signals were cut off and reappeared, and thus could calculate an exact position. Optical astronomers found it to be a fairly bright 'star'– visible with a backyard telescope – and again with a peculiar

spectrum. In February 1963, Dutch–American astronomer Maarten Schmidt suddenly hit on the answer. In one of those rare flashes of insight which are the high points of a scientific career, he realized that the pattern of lines was simply the 'fingerprint' of hydrogen – but shifted towards the red end of the spectrum by 15·8%. If this was a Doppler shift, the starlike object was travelling away from us at 45,000 kilometres (27,900 miles) per second, far too fast for it to be a star of the Milky Way's family. Since they were evidently not stars, these objects were entitled 'quasi-stellar objects', later conveniently shortened to 'quasars'.

There's one natural interpretation. As we'll see in more detail in the next chapter, the Universe is expanding; so that the further away a galaxy is, the faster it's moving away from us, and the more its spectral lines are Doppler-shifted towards the red end of the spectrum. The quasars could be far away in the realm of the galaxies, moving away with the general expansion of the Universe; and the wavelength shift would simply be telling us their distance. 3C 273 would then be 2,000 million light years away from our Galaxy.

But in that case, quasars are immensely bright. Many quasars turn out to be a hundred times brighter than normal galaxies, and yet they are comparatively minute. They appear as no more than points of light on photographic plates; 'starlike', rather than the extended blur of a distant galaxy. And astronomers can use a simple argument to say roughly what size a quasar is, along the following lines.

Imagine that you could suddenly switch off the Sun. Now, the rim of the Sun is slightly farther from the Earth than is the centre of the disc, and so light signals from the

The 11 m (36 foot) diameter
telescope at Kitt Peak National
Observatory, Arizona, detects
radiation on the borderline
between radio and infra-red, and

so it is a cross between an
ordinary reflecting telescope and a
radio telescope. These wavelengths
are emitted strongly by quasars.
Courtesy Ian Robson.

edge take about two seconds longer to reach us. Although the Sun's surface all goes dark in an instant, at the time we see the centre turn black, our eyes are still picking up photons from the edge, which left two seconds earlier, when the Sun was still bright. So we'd see a wave of darkening spreading from the centre outwards, to reach the Sun's rim two seconds later. The Sun's light would seem to switch off slowly, simply because of its size. A distant astronomer, who saw the Sun merely as a point of light, would be able to tell that our Sun is about two light seconds in radius simply by watching it fade.

Quasars are by no means constant beacons in space. They 'flicker', brightening and fading as erratically as a candle in a breeze, but their flickering takes place on a scale of months. So astronomers can immediately say that quasars are only a few light months across. And here was the crunch. If quasars are as far away as their Doppler shifts indicated, then they are producing as much light as a hundred galaxies from a region nearly a million times smaller than a galaxy. That's as much light concentrated together as packing ten million million stars into a region of space one-tenth the size of the distance from the Sun to the nearest star!

Some astronomers not unnaturally baulked, and the 1960s and early '70s saw the great quasar controversy. Astronomers of high standing confronted one another and disputed just what quasars are, how far away they are and what it is that shifts their spectral lines to the red. Some astronomers went so far as to claim that new laws of physics had to be invented to explain the quasars. Yet the controversy has quieted down now: most astronomers are coming to agree that quasars really do pack that much power into such a small volume; and that they really are the most distant objects in the Universe.

Quasars were discovered ahead of their time. Astronomers of the early 1960s weren't prepared to accept such unbridled violence in their orderly Universe. Since then, all branches of astronomy have been highlighting the profligate energy that Nature casually lets loose. On a small scale (in the galactic sense), X-ray astronomers have discovered the powerful X-ray binary sources, where gas falling on to a neutron star has some of its mass converted into energy; and the even more dramatic black holes, where the surrounding gas spiralling inwards has nearly half its mass turned into energy. In 1963, in fact, the ultra-compact neutron stars and the still more exotic black holes were unknown. No one then realized that Nature has provided ways to convert matter so efficiently into energy, and so the energy source of quasars was naturally an enigma.

Since then, astronomers studying the extragalactic scene have come to realize that the Universe has a whole range of violence. There are not just quiet, ordinary galaxies at one extreme, and quasars at the other. Most galaxies seem to have a tiny central 'nucleus', and this seems to be the key to the quasar phenomenon. Some galaxies' nuclei are effectively 'mini-quasars', for example; and we can see the whole gamut of galaxies, from those whose nucleus is too feeble to shine or affect the surrounding galaxy, through those where it shines weakly and is blowing clouds of gas from the centre, right up to those whose nucleus is so bright that it's difficult to see the surrounding galaxy at all. The 'classical' quasars seem to be simply the most energetic nuclei of all, the most violent criminals in a Universe where all galaxies' nuclei are to a greater or lesser extent thugs. And because quasars are so brash, astronomers naturally noticed them before their tamer brothers, and published their exploits in the astronomy journals.

The mini-quasars

Even our own Galaxy seems to harbour a 'sub-mini-quasar', or 'micro-quasar', in its heart. The centre of our Galaxy was unknown territory until recent years, for optical astronomers can't study it at all; dust grains between it and us absorb practically all the light. It's estimated that only one part in a hundred thousand million of the light from the galactic centre gets through to us – and that's not enough to detect, with even the largest telescopes. So optical astronomers are in the odd position of knowing more about the nuclei of other galaxies millions of light years away than about the nucleus of our own Galaxy.

But longer-wavelength radiation can penetrate the dust, and radio astronomers have known for twenty years that huge gas clouds near the galactic centre are racing outwards, as if from an explosion several million years ago. Now, with improved techniques, they have homed in on the central regions, to find a small radio source, only as big as the orbit of Jupiter around the Sun. On the scale of our Galaxy, this is a minuscule volume of space; and in comparison to its size it is a tremendously powerful radio transmitter: volume for volume, it is as strong as the radio-emitting cores of quasars.

Infra-red astronomers have filled in another piece of the puzzle. They see the infra-red brightness increasing towards the Galaxy's centre, simply because the stars are becoming more closely packed. If we lived within the central one-light-year-across region of the Galaxy, the sky would be studded with millions of bright nearby stars, the brightest of them rivalling the full Moon. There would be no real night, for the stars would keep the sky light all around the clock.

The space between the central stars is threaded by gas streamers, and the movements of this gas tells infra-red astronomers just how much mass there is. In 1977, such measurements showed that there seems to be a lot more mass near the centre than the stars can account for. 'Missing mass' again, but this time right at our Galaxy's heart. Although these difficult measurements need to be checked, some astronomers have taken the plunge and suggest that this extra mass forms a giant black hole at the centre of the Galaxy. This massive maw, five million times heavier than the Sun (but only twenty times larger) would not just explain the missing mass; gas and broken-up stars could spiral down into it and their mass could be converted into the huge amounts of radio waves that we see from the galactic centre. The radio source constituting the 'micro-quasar' could be the surrounding region of a massive black hole which marks the precise centre of our Milky Way Galaxy.

We may seem to have entered the realm of science fiction here, with huge black holes swallowing stars. But there are sound reasons for believing that black holes are the 'best-buy' explanation for mini-quasars and quasars – to quote Cambridge astronomer Martin Rees. Let's look at some of the evidence: behaviour so bizarre that it's difficult to explain in any other way.

Other spiral galaxies may have micro-quasars, like ours, but most would be too faint to pick up. The next step up the mini-quasar scale are the Seyfert galaxies. First noted by American astronomer Carl Seyfert in 1943, these are apparently normal spiral galaxies, but with a central, exceptionally bright, starlike nucleus – what we could now call a mini-quasar. Around the centre, gas streamers and clouds are being blown outwards at enormous speeds: thousands of kilometres per second. Clearly the mini-quasar marks the site of some kind of tremendous explosion.

Interestingly enough, about one in ten of the large spiral galaxies – the same type as the Milky Way – has one of these bright nuclei and exploding gas clouds. And there are good reasons for thinking that the Seyfert galaxies we see haven't always been so active. It's more likely that every so often their nuclei go through a flare-up. And that raises the fascinating possibility that all large spirals may flare up occasionally at the nucleus, and appear as mini-quasar Seyfert galaxies. Perhaps the micro-quasar at our own Galaxy's heart is now just resting: in the past, and again in the future, it may flare up into the mini-quasar activity of a Seyfert's nucleus.

When we investigate the spectrum of a Seyfert's centre,

we find that the 'mini-quasar' name is even more apt. The nucleus is intrinsically fainter than a quasar, and the spectral lines are less Doppler-shifted in wavelength, simply because the Seyferts are comparatively nearby on the cosmic scale – but otherwise, the spectrum is almost identical to a quasar's. Whatever is going on in a quasar is also happening on a smaller scale in the Seyfert nucleus.

These nuclei also flicker in brightness like a quasar. Ironically enough, some of the brighter-nuclei Seyfert galaxies were first classified as variable stars, for on short-exposure photographs the surrounding 'fuzz' of the stars making up the galaxy didn't show, and only the bright, flickering 'starlike' nucleus could be seen. Since this was before the days when astronomers had accepted violence on a galactic scale, the varying, 'starlike' points of light were naturally assumed to be faint variable stars in our own Galaxy.

In 1978, astronomers discovered a relatively nearby Seyfert galaxy whose nucleus is as bright and powerful as a real quasar. Catalogued ESO 113-IG45, or Fairall 9 after its discoverer, South African astronomer Anthony Fairall, this galaxy is the 'missing link' between Seyferts and quasars proper. It's close enough for us to see both the quasar-like centre and the surrounding galaxy of stars. If we could take Fairall 9 farther and farther away, the surrounding galaxy, appearing spread-out and fuzzy, would become invisible first, while the bright point of the nucleus would still show up on photographs. At this distance, the galaxy would be moving quite fast with the expansion of the Universe, and the spectrum of the nucleus would have a large Doppler shift: astronomers would say 'this is a quasar'.

American astronomer Jerome Kristian has tackled the same question from the other end, to find the same answer. He selected the nearest and intrinsically weakest quasars, and discovered that long-exposure photographs do show up a faint surrounding 'fuzz' – the combined light of all the ordinary stars in the galaxy surrounding the quasar nucleus.

So quasars are not such strange, utterly different objects as they appeared at first. They are just super-powerful versions of the nuclei of galaxies, differing only in degree of violence from those which Carl Seyfert started studying in the 1940s. As we might expect, these extremely violent galaxy nuclei are fairly rare in the Universe, so most of the powerful ones are a long way off. Only in the very few nearby ones, like Fairall 9, can we see the surrounding galaxy. Yet the immense power of these nuclei, making the central regions shine so brightly (quasar 3C 279, for example, can flare to ten thousand times brighter than our entire galaxy) means that astronomers can see them even at enormous distances. Astronomers can pick out quasars far beyond any ordinary galaxy; these celestial beacons are the farthest objects which telescopes can show us.

Messages from the radio galaxies

The quasars are undoubtedly spectacular. But they give away very little about where their stupendous energy output originates. Large black holes are certainly a possibility, but astronomers need more evidence before dismissing other alternatives. And here radio astronomers have unearthed a vital clue, in galaxies which don't look so unusual on photographs, but are powerful radio transmitters.

Cygnus A was the first powerful radio galaxy to be identified. As we've seen already, astronomers were taken aback by the power of its radio transmission, and the shock was compounded when more sensitive measurements showed that the radio waves weren't coming from the centre of the galaxy. Instead, there are huge clouds, one on either side of the galaxy, which are generating the radio waves. Cygnus A has turned out to be typical of other radio galaxies (hundreds are now known), for practically all these galaxies are straddled by a pair of radio-broadcasting clouds. If our eyes were sensitive to radio waves, we would literally be seeing double at night, for the galaxies themselves would be invisible, and we would see the sky (neglecting radio sources in our Galaxy) full of double 'clouds' instead.

The great Russian astrophysicist Josef Shklovsky calculated early on that the type of radio emission from these clouds must arise from very fast-moving electrons travelling in a magnetic field. The problem which has plagued radio astronomers for the last twenty years is how the energetic electrons and magnetic field come to be in the virtually empty space outside the galaxy; and why they should be clumped into two distinct clouds, astride the galaxy.

A pair of large radio-emitting clouds may not seem as violent an event as the tightly packed activity in a quasar nucleus, but calculations show that an enormous amount of energy is involved. For a start, the radio clouds in Cygnus A are broadcasting as strongly in radio waves as the Milky Way Galaxy does in the form of light from all its stars; and yet these radio-emitting clouds are in practically empty space. Further calculations show that the actual energy supply locked up in the clouds – as magnetic field and the electrons' energy of motion – is enormous beyond belief. Let's put it this way. To keep the Sun shining for its entire lifetime, about one-thousandth of its mass will be turned to energy, according to Einstein's $E = mc^2$ formula. Think of the Sun's continuous output of light and heat, and multiply that by ten thousand million years: that's the amount of energy equivalent to just one-thousandth the Sun's mass. Each of the radio-emitting clouds of Cygnus A contains the energy equivalent of a hundred thousand Suns. In other words, you would have to destroy this many typical stars completely and turn them completely to energy in the form of fast electrons and magnetic field to produce such a powerful radio source. And yet these clouds are outside

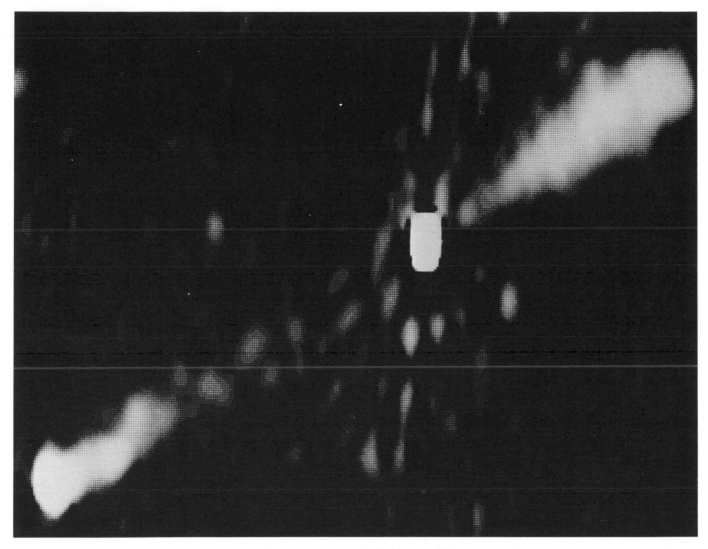

the galaxy itself, where there are no stars to provide any energy at all.

Without a doubt, the energy must come from the galaxy's nucleus. Some quasars have a very similar pair of large radio clouds, and here the energy source is obviously the violent nucleus. Many radio galaxies, too, have a tiny radio source at their hearts, showing that something special is going on, even if it doesn't cause the nucleus to shine as brightly as a quasar.

The most popular explanation among radio astronomers today is that the galaxy's nucleus shoots out 'beams' of fast electrons in opposite directions. These travel outwards, blasting tunnels for themselves through the very thin gas lying around the galaxy; and, at the far end of the ever-growing tunnel, the electrons, which have been moving untrammelled up to now, crash into the surrounding gas. Their motion is disrupted; they get tied up in a tangled magnetic field (which has either moved out with them, or is

generated by the electrons' turbulent motion) and their energy is radiated away as radio waves. Electrons splashed back from the collision with surrounding gas spread out into a huge, radio-transmitting cloud; while the tunnel end itself should appear as a particularly powerful radio 'hot spot'.

Powerful radio galaxies do indeed have 'hot spots' at the outer edges of the clouds. And in 1977, astronomers succeeded in finding faint light coming from just these regions. Like the radio waves, the light comes from electrons twirling around in a magnetic field after rebounding from the stationary gas at the end of the tunnel. There's as much light coming from these ethereal collisions as from the millions of stars in a small normal galaxy. An astronomer living on a planet in a radio galaxy would see faint companion 'galaxies', exactly opposite each other in the sky; but these would not be ordinary galaxies of stars, like the Milky Way's Magellanic Cloud companions. They would

appear as eerie, shapeless, misty patches, revealing nothing more in the telescope than a bland, white glow, like splashes of luminous paint on the black bowl of night. Incidentally, there might well be more people on such a planet watching the evening sky, for the intense radio noise coming from the 'clouds' would completely jam any television transmissions!

Radio astronomers have looked hard for signs of the tunnels through which the electrons move, and some galaxies do show such faint radio connections from nucleus to outer clouds. Some of the tunnels even shine in ordinary light. Photographs of the giant elliptical galaxy Messier 87 show a narrow, blobby streak stretching out from its centre; and the famous quasar 3C 273 has an even longer 'jet', extending out over 200,000 light years.

Once again, we need a powerful nucleus. The energy of radio galaxies comes out from the nucleus, and it is here that the greater part of the mass of a million stars must be transformed into energy, before being beamed outwards in the form of fast electrons. Many radio galaxies have suffered more than one outburst. As well as the huge clouds far out from the galaxy, there is a smaller pair close in. The beam seems to have switched off for a while; and, when it has lit up again, it has had to start tunnelling its way out through the surrounding gas again. By simultaneously using radio telescopes on different continents to give the effect of one very large telescope, radio astronomers have probed the details of these small new radio clouds, and have actually seen them expanding outwards from the central few light years. Whatever is producing the energy is as concentrated as the quasars' energy source. And, most surprisingly, this second pair of radio clouds is always along exactly the same line as the original outer pair. Somehow the beam has 'remembered' its exact direction when it has switched on again.

Powerhouse in the quasar?

All these results can be tied together neatly, if we assume that lurking in the centre of each radio galaxy and quasar there is a massive black hole, roughly a thousand million times heavier than the Sun, although smaller than the solar system in diameter. (To be fair, this isn't the only possible explanation, but it seems to fit the facts best.) As stars orbiting near the centre of such a galaxy come close to the black hole, they get broken up. This isn't because the black hole in some way 'sucks them in'; as we saw in Chapter 9, from a distance a black hole's gravitational pull is the same as any other massive body's. Stars may collide with one another in the dense star cluster which forms the galaxy's centre, and they will break open to spill their interiors as streams of hot gas. Stars which stray too near the black hole find that they are stretched by tides – just as the Moon's pull distorts the Earth slightly – until they are literally stretched to breaking point, and are ripped apart. All this gas from broken-up stars spirals inwards towards the black hole, and builds up into a huge, rotating disc of hot gas around the hole. Radiation from this searingly hot disc can actually rip away the outer layers of giant stars which come too close, and this plunder will also join the gaseous disc around the black hole.

Gas spiralling into a black hole produces copious amounts of energy, as we noted before. A greedy black hole can generate the energy for quasars and radio galaxies in the most efficient way known, and the small size of the energy-producing disc ties in well with the tininess of quasar and radio-galaxy cores. Calculations show that such a black hole would only have to swallow gas equivalent to one Sun-like star per year to keep up the power output; and this is quite reasonable for a black hole finding itself in the centre of a giant galaxy. Astronomers still dispute exactly how this energy output can be converted to light, but it's easy to explain why quasars 'flicker', simply as a result of the varying amounts of gas in the spiralling accretion disc.

The whirling magnetized disc could act as a generator, too, speeding up electrons to near-light velocities, and shooting them away at right angles, up and down the axis of rotation. This could be the origin of the double beam which carries the energy out into the broadcasting clouds of a radio galaxy. It certainly explains the directional 'memory', for a rotation axis is about the only thing in astronomy which always points in the same direction. If the disc dies down through lack of gas, the beams of electrons switch off; but as soon as fresh gas supplies create another disc, the new beams must follow the same paths as before, along the axis of rotation.

Our Galaxy contains a micro-quasar, possibly an active region around a five-million-Sun black hole; quasars and radio galaxies may harbour black holes a thousand times heavier in their nuclei. Astrophysicists can explain roughly how active galactic nuclei, from our micro-quasar, through the central mini-quasars of the Seyfert galaxies, to radio galaxies and quasars themselves, can result from varying amounts of gas spiralling into large black holes. The details need to be worked out, but there's a feeling among astronomers that at last a good explanation has been found for the violence at the centres of galaxies.

To test this picture, can we find a way to look for such black holes more directly? By definition, we cannot actually see a black hole; but its gravitational pull may affect the motions of stars in quite a large region around. Following this idea, a team of Californian astronomers cracked down on the nearby giant elliptical galaxy M87 in 1977. Using the latest electronic image devices on large telescopes, they looked in detail at its centre. The stars here were indeed more closely packed than they should be, as if a massive central object were holding them in; and these stars were moving faster than expected. To keep these high-speed stars in their orbits, again there must be a strong gravitational pull. Careful measurements showed that both the close packing and the high speeds indicated a central, unseen body, containing as much matter as several thousand million Suns, and certainly less than a few hundred light years across. The only object which fits the bill is a massive black hole. And it's probably no coincidence that M87 is a radio galaxy, with its electron tunnel visible as the famous optical 'jet'.

Forty years ago, astronomers saw galaxies as placid Catherine wheels and balls of stars, rotating peacefully in the depths of space. Our modern picture is quite different, for galaxies have hidden secrets which only the most up-to-date research methods can discover. The outstanding problems now crying out for solution are whether most of the matter in galaxies is actually invisible; and just what does power the quasars. Astronomers are faced with the task of either finding out what the 'missing mass' in galaxy haloes and clusters of galaxies is, or else explaining away the embarrassing results in some other way. And although a massive black hole seems the best explanation for the power of active nuclei, right up to the scale of quasars, there may be a better answer. Whatever that may be, it would undoubtedly have to be even stranger than black holes.

Finally, astronomers are still uncertain about how the galaxies formed in the first place. Above, we've traced a general picture for our own Galaxy, but even here they are far from sure of the details. We've reached the stage where the visible and radio-emitting contents of galaxies – their stars and gas clouds – are succumbing to the intensive investigation of modern techniques; but, looked at on a wider scale, galaxies still hold secrets in store. Indeed, the world of galaxy research is one of the fastest-moving in all astronomy. Half this chapter could not have been written ten years ago; and few astronomers would care to bet on what the next ten years may hold.

These lava outflows built up the new volcanic island of Surtsey, off Iceland, in 1963. Change occurs everywhere in the Universe; the passage of time is irresistible. It is possible that time itself had a beginning, and may eventually end.

11:Beginning and End of Time

A medieval view of the Universe:
the heavens are a large dome,
whose motions are controlled
by some outside mechanism. This

woodcut, although demonstrating
earlier ideas, dates from the
nineteenth century.

The science of *cosmology* began when Man first looked up at the heavens and asked: 'Where has it all come from?' Earth and Heaven were born from chaos, according to many ancient myths, and until recent years myths and religious dogma were our only guides to the birth of the Universe. The seventeenth century Bishop Ussher of Armagh, for example, calculated the exact date of Creation by adding up the ages of the biblical patriarchs, and concluded that the first day began in 4004 BC.

Scientists were slower off the mark to make a stab at the age of the Universe. Well into this century, only geologists had good hard evidence. For the thick layers of sedimentary rocks laid down at microscopically slow rates, they needed at least several hundred million years of past history. As we have seen, the radioactive dating of rocks from the Earth, Moon and meteorites has now fixed the formation of our

system fairly precisely to 4,600 million years ago. Clearly, the Universe must be older than this.

But in the last twenty years, astronomers have amassed a surprising amount of data, which pinpoint not just when the Universe began, but also how. By studying clusters of stars, the elements in space, the motions of distant galaxies and the faint radio 'noises' from the depths of the Universe, present-day cosmologists find that a surprisingly simple and consistent picture is emerging. And these observations will also allow us to calculate what the Universe's future will be. Studying the beginning and end of the Universe – and of space and time, too, in the modern picture – is no longer the province of theologians and philosophers. Cosmologists are now scientists, with firm facts to work on, and theories which can be tested by observation, as well as confirmed by scientists in other fields.

The mysterious dark sky

There's one very important cosmological observation that we've all made since childhood: the night sky (between the stars) is black. That doesn't sound very fundamental. Indeed it may seem obvious that the sky should be dark at night; but is our intuition correct?

English astronomer Edmund Halley (1656–1742) seems to have been the first to ponder on the blackness of the sky, although the effect is usually called Olbers' paradox, after another comet seeker, the German Heinrich Olbers (1758–1840). Let's suppose that the Universe is infinite: it stretches on for ever and ever and it is filled with stars throughout. Now, wherever we look in the sky, we should see a star's surface. We may see between the nearby stars, because their sizes are so small compared to the distances between them; but somewhere our line of sight – stretching out through an infinite Universe – must meet a star. So every spot in the sky should be at the temperature of a star's surface, shining down on us with a temperature of thousands of degrees!

You might argue that the more distant stars would be fainter. But there are more of them, and the two effects exactly cancel. And the clumping of galaxies has no effect, as long as galaxies of stars continue out indefinitely into space. The darkness of the sky is proof that our Universe isn't just made of galaxies stretching out throughout infinite space, sitting in place since time immemorial.

Obviously we can escape from the dilemma by saying that the realm of galaxies has an edge, beyond which there is nothing. But there are two other escape routes. If we remember that light travels at a definite speed, then looking out into space is also looking back in time. The light by which we see the Andromeda Galaxy left it two million years ago; light and radio waves that we pick up from the disturbed galaxy Cygnus A began their journey 700 million years ago. If all the galaxies came into being at some particular time in the past, there must be a certain distance corresponding to that time, and anything coming from beyond this distance will date from a time before any galaxies existed.

As we found in the last chapter, astronomers now believe that galaxies were born some 15,000 million years ago. Anything in the Universe further away than 15,000 million light years will appear to us the way it was before galaxies formed. The snag is that before galaxy formation, the gas filling the Universe was as hot as star surfaces, so we can't take this escape route from Olbers' paradox.

But the night sky would be dark if the Universe were simply expanding. Then the light from the distant, receding galaxies would have its wavelength changed by the Doppler effect. All wavelengths would be stretched, turning the initially visible light from stars into longer infra-red or radio waves; and so the galaxies' stars would no longer be visible.

The most distant stars would thus be effectively black. Indeed, if we think of an individual photon of light from a distant galaxy, its increasing wavelength means a correspondingly lower energy. So the total energy we pick up from a distant galaxy, including all wavelengths of radiation, decreases as well.

Astronomers now know that the Universe is expanding. Although galaxies and their contents stay the same size, and clusters of galaxies remain tied together by their gravitational attraction, every cluster is racing away from every other cluster. Observing the details of this expansion is a tedious and time-consuming occupation, but it's fascinating that the possibility of an expanding Universe could have been predicted centuries ago from the blackness of the night sky: Oxford cosmologist Dennis Sciama calls this the greatest missed opportunity in the history of cosmology.

The expanding Universe

Many astronomers would rate the expansion of the Universe as the most important single fact we know about it; but the man who discovered the remarkable motion of the galaxies was not obviously destined for astronomical fame. Edwin Hubble (1889–1953) began his career as a lawyer in Louisville, Kentucky. But at the age of 25, to use his own words, he 'chucked the law for astronomy, and I knew that even if I were second-rate or third-rate it was astronomy which mattered'. Hubble turned out to be no second-rate astronomer. Armed with the then largest telescope in the world, the 100-inch Mount Wilson reflector, he became the leading extragalactic astronomer of his generation.

His greatest discovery followed from a series of long and difficult observations by his assistant Milton Humason. (Humason himself entered astronomy through an unusual channel: he was initially employed as a janitor at the Mount Wilson Observatory.) They were investigating the motions of galaxies by looking for the Doppler shifts in their spectral lines, which would show how fast the galaxies are travelling towards or away from us. Even with a large telescope, this is not an easy task; for a galaxy's light comes from an extended blur in the sky, and it's a problem to get sufficient light to expose a photographic plate, especially when the available light has to be spread out along the spectrum. To obtain a spectrum of one of his faintest and most distant galaxies, Humason exposed the same photographic plate on five successive nights, to give a total exposure time of 45 hours. Allan Sandage, one of Hubble's students, and now a leading American astronomer himself, points out that a modern electronic device on the same telescope can now get as good a spectrum in only ten minutes.

Astronomers had known for some years that most galaxies are racing away from our own Milky Way Galaxy. (One

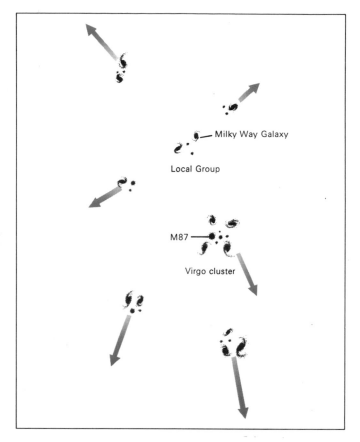

Milky Way Galaxy

Local Group

M87

Virgo cluster

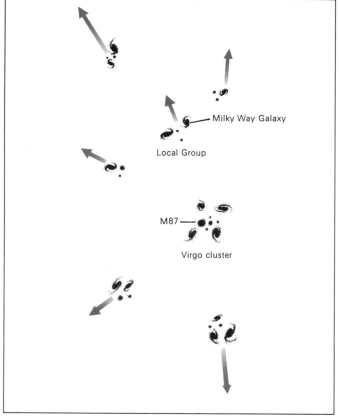

Milky Way Galaxy

Local Group

M87

Virgo cluster

exception in fact is the Andromeda Galaxy, which appears to be speeding towards us; but we now know that this is mainly due to the Sun's orbit around the centre of our Galaxy, which is at present carrying us rapidly in the direction of Andromeda.) Hubble discovered that a galaxy's speed away from us depends on its distance; and that speed increases in step with distance as we look at more and more remote galaxies. If one galaxy is twice as far away as another, it will be racing away at twice the speed.

This simple rule which governs the galaxies' motions is now called 'Hubble's Law'. Galaxies' spectra can now be measured out to fifty times the distance of Humason's furthest galaxy, and their velocities still increase in step. The farthest galaxy which astronomers now know is speeding away at about half the speed of light, and the even more distant quasars travel even faster.

We can make a useful analogy using aircraft. Suppose two aircraft take off from New York heading for London, the first at 500 kilometres per hour, the second at 1,000 kilometres per hour. In one hour, the first will be 500 kilometres closer to London (or away from New York), while the other will be twice as far from New York, or twice the distance closer to London. If we know their relative positions and speeds, we can calculate the time of take-off. And so it is on the cosmic scale.

All the galaxies must have been extremely closely packed together in the past, and they have sped apart to different distances because of their various speeds. But to calculate the time of the celestial 'take-off', we must know the galaxies' distances as well as their speeds; and this was Hubble's problem. He knew the relative distances of his galaxies, simply because the farther ones seem fainter. How far away they are is a question which still isn't answered exactly.

Hubble's best estimate for galaxy distances put the galaxies' 'take-off' to a moment 2,000 million years ago. Before this time they must have been so tightly packed that they wouldn't have been recognizable as galaxies; indeed it must have been before the time of galaxy formation. But even when Hubble made this estimate, the first radioactive measurements of the Earth's age had shown it to be around 3,000 million years old. The Earth could hardly be older than the galaxies.

Astronomers have plugged away at refining distance measurements since then, however, and gradually they have found that the galaxies are considerably further away than Hubble thought. There's still some uncertainty, indeed. Many astronomers would increase Hubble's distances five times, while Hubble's former student Allan Sandage believes that an increase of ten times would be nearer the mark. Certainly the galaxies are much more remote than the early work indicated, and so their take-off must have happened five to ten times longer ago – say 15,000 million years as a convenient figure. This would make the galaxies older than the Earth, and it agrees very nicely with the age of our Galaxy, which astronomers have calculated from old star clusters (in Chapter 10).

Another estimate of our Galaxy's age comes from a type of radioactive dating on some of the heavy elements made in supernovae (a method with the unwieldy name of nucleo-cosmochronology). As soon as our Galaxy formed, heavy stars began exploding as supernovae, pouring out heavy, unstable elements into space. Looking at the present-day quantities of different isotopes of elements such as the rare element rhenium, astrophysicists reckon that our Galaxy's first supernovae appeared between ten and twenty thousand million years ago.

Astronomers are now confident that the 'age of the Universe' lies between these limits, and the figure of 15,000 million years cannot be far from the truth. What happened at that time we'll see a little later, but there's one thing about Hubble's Law which may still bother us. From the fact that all the galaxies are racing away, it might seem that our Galaxy must be in the centre of the Universe – and that, certainly, would be a remarkable coincidence.

The explanation lies in another result of Hubble's simple law. Suppose we lived in another, distant, galaxy, one which Earthly astronomers see fleeing away at 1,000 kilometres per second – say, the giant elliptical galaxy M87. We would then think our home galaxy M87 to be at rest: relative to us, the Milky Way Galaxy would seem to be shooting off at 1,000 kilometres per second in the other direction. And indeed, if we calculated how the other galaxies in the Universe would move when seen from our new home in M87, we find that Hubble's Law is true from there, too. Although the speeds of all the galaxies are now different, relative to us, when measured from M87, their distances are also changed when we measure them from there; and the two effects cancel each other out. Galactic velocities increase with distance for the M87 astronomer exactly as they do for a Milky Way astronomer.

In the aircraft analogy, we could imagine ourselves travelling on the slower aircraft at 500 kilometres per hour. But our speed is only relative: suppose it's dark out so that we cannot see the ground. If we imagine ourselves to be standing still, then the faster aircraft is speeding away from us at 500 kilometres per hour – and New York is travelling away in the opposite direction at the same speed. In the Universe, all motion is relative; 'New York' is just another 'aircraft'.

Astronomers are very happy that the Milky Way doesn't turn out to be in a specially privileged place in the Universe.

Ever since Copernicus dethroned the Earth from the centre of the Universe, and showed it to be simply a planet circling the Sun, scientists have been wary of ascribing it a special place. Since Copernicus' time, they have found that the Sun is merely a very average star, circling two-thirds of the way out in an average galaxy. And now it seems that our Galaxy's place in the Universe is nothing unusual.

Looking at it in another way, we can say that if our position in the Universe is typical, then the view from any other galaxy must look much the same. Obviously, it may have different kinds of neighbouring galaxies; but once we look out beyond the nearest galaxies and clusters of galaxies, the distant parts of the Universe should appear very similar to our view from the Milky Way. In the days when cosmologists had very few facts, and many theories, this idea was a very important guideline, and it was enshrined as the Cosmological Principle. Recent results haven't contradicted this principle, which is still very important to theorists working on detailed calculations involving the Universe as a whole.

Modern cosmologists see the Universe as a huge arena in which all the galaxies are flying apart from each other, in such a way that any galaxy can consider itself to be the centre of the Universe. But you might think there must be a 'real centre', the point where the original explosion (or take-off) which dispersed the galaxies occurred.

As it happens, there isn't such a centre. When we deal with the Universe, working on the largest possible scale of things we know, it turns out that we must to some extent take leave of common sense, just as we did when exploring the very small in Chapter 6. Here we must rely on mathematical theories, which may seem strange at first, but which do – unlike common sense – agree with the experiments and observations. For the very big, cosmologists use Einstein's General Theory of Relativity, the theory which deals with gravitation, space and time.

Applying the theory, we find that the Universe didn't just begin with an explosion at one point in space, a 'shrapnel burst' which threw out the galaxies. The theory tells us that space itself is expanding, and carrying the clusters of galaxies willy-nilly, like boats in a rip tide. Inside a galaxy, the stars' gravitational attraction keeps them bound as a single unit, while the concentrated clusters of galaxies again bring in a localized gravitational force which holds each cluster together. But neighbouring clusters have little gravitational hold on each other. The empty space between them expands continuously, and the clusters are carried helplessly apart. The idea of empty space 'expanding' may bemuse the mind; but in Einstein's theory, space has a real existence, and expansion of space must be taken absolutely seriously.

We could make an analogy with a fruit cake in the oven.

A cutaway view of the entire
visible Universe. From our
Galaxy (centre), looking outwards
in space is looking backwards in
time, since light does not travel
instantaneously. Sufficiently far

away, we are seeing the Universe
in its infancy, when galaxies were
closely packed and quasars
common. At a distance of
15,000 million light years, we
are looking back to the beginning

of the Universe itself, the Big
Bang, which we 'see' in all
directions around us. (Note that
this does not mean we are at
some special central point in the

Universe: the same diagram
could be drawn centred on any
other galaxy.) The enlarged
segment shows the events which
occurred soon after the Big Bang.

As it rises, the raisins are taken apart from each other, not because of any explosion, but because the cake between them is growing. For 'raisins' read 'clusters of galaxies', and for 'cake' substitute 'empty space', and you'll have some idea of what is going on in the expanding Universe.

But fruit cakes have edges. And since astronomers don't believe that the Universe has an edge, the analogy is not exact. The theory shows that two types of Universe are possible, in fact. In one type, the clusters of galaxies just go on forever. The Universe is infinite, but we don't see it all because it isn't infinitely old. As we discovered in Olbers' paradox, we can only see out to 15,000 million light years, for at that distance we are looking right back to the beginning of the Universe. That's what cosmologists call an 'open' Universe.

A closed Universe is even more bizarre. If we set out in a straight line from the Milky Way and travelled for a very long time, we would eventually find ourselves approaching the Milky Way again from the other direction! Space has bent around on itself. It's slightly like travelling around the Earth, setting off in one direction and finding yourself returning home from the opposite. This analogy is easy to understand – unless we believe that the Earth is flat. And when Man first began to understand that the Earth was round, it must have been a little hard to imagine that gravity makes everything seem right side up, and that people in China are not walking around upside down. The snag is that the bending around of the Universe is in a fourth dimension – and that we can't just visualize. We couldn't see all the way round such a Universe, incidentally, for the total 'round trip' (in a straight line, remember) from our Galaxy to our Galaxy again is longer than 15,000 million light years. Again, we see out just as far as 15,000 million light years in every direction, for at this distance we have got back to the beginning of the Universe.

Let's stick with the open Universe for the moment, though, since it's slightly easier to comprehend. We can see now why astronomers' least favourite question is 'how big is the Universe?' The Universe may well be infinite, but we can't see all of it – and the reasons given above aren't the kind you can explain in two minutes in words of one syllable.

When astronomers say that the Universe is expanding, what they really mean is that the distances between the clusters of galaxies is expanding. (Architects, landscape gardeners and other artists, by the way, are also aware that space between objects is as important as the objects themselves.) The important thing for us to try to understand is that the Universe – that is, Space – has always been infinite: 7,500 million years ago, the clusters of galaxies were half as far apart as they are now, but half of infinity is still infinity. (This is slightly different for a closed universe, but that's even more complicated.)

There are two ways of talking about the history of the Universe. Theorists are quite happy to think of the lump of gas which started where our Galaxy is now, and follow through its history to the present day. Alternatively, we can look out into space, and use the speed of light. We see Cygnus A as it was 700 million years ago; 3C 273 in the state it was in 2,000 million years ago – that is, 13,000 million years after the Universe's beginning. (Since this was the moment when space began to expand, dragging matter with it, it's been nicknamed the 'Big Bang', even though it wasn't an explosion in the normal sense.)

Looking out at the most distant quasars, astronomers are seeing nine-tenths of the way back in time. Even allowing for the effect of the expansion of space, there are many more quasars at these distances than there are in the 'recent' region of space near the Milky Way Galaxy. It seems that quasars were much more common in the Universe's early life, and their violence is a symptom of youth. In fact, many astronomers expect this, for in the initial early throes of a galaxy's life there will be a lot of gas around which could fall into a central black hole, and make a violent quasar at the galaxy's heart.

Beyond the quasars – and earlier than them in time – we might expect to see galaxies forming. So far, astronomers have not succeeded in picking up light from these 'proto-galaxies', despite intensive searches. Farther on still, what should we expect?

The remarkable Russian-American physicist George Gamow – well known for his popular science books starring 'Mr Tompkins' – came up with the right answer thirty years ago, when cosmology was all but a black art to other scientists. He reasoned that in its early stages the Universe must have contained hot gas and radiation, and that this radiation should have been lengthened by the Doppler shift during its journey across thousands of millions of light years of expanding space which it must traverse to reach us. It should reach us as radio waves. But at that time radio telescopes were not sensitive enough to detect such a signal from the depths of the Universe, and Gamow's prediction was lost – until the incident of the 'noisy' radio antenna at a New Jersey satellite-tracking station.

Whispers from the birth of the Universe

Bell Telephone Laboratories have already cropped up in our story, for it was one of their engineers who – by accident – first detected radio waves from space in the 1930s. Their huge satellite-tracking antenna at Holmdel, New Jersey, was designed to be one of the most sensitive antennas in the world, at its working wavelength of a few centimetres. For this reason, radio astronomers Arno Penzias and Robert Wilson 'borrowed' it for a while in 1965

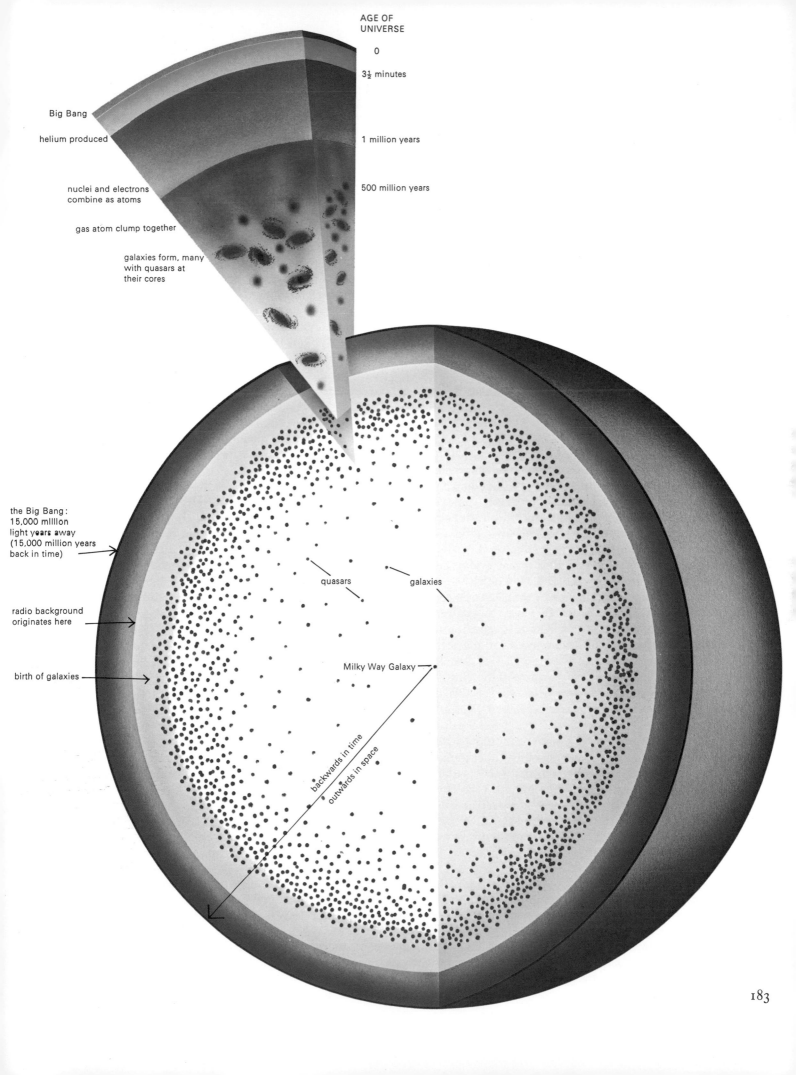

AGE OF
UNIVERSE

0

3½ minutes

1 million years

500 million years

Big Bang

helium produced

nuclei and electrons
combine as atoms

gas atom clump together

galaxies form, many
with quasars at
their cores

the Big Bang:
15,000 million
light years away
(15,000 million years
back in time)

radio background
originates here

birth of galaxies

quasars   galaxies

Milky Way Galaxy

backwards in time
outwards in space

Opposite page, top:
In an experiment flown in a
modified U-2 aircraft in 1977,
George F. Smoot (right
foreground) and colleagues
studied the 'brightness' of the

radio background lingering from
the Big Bang. The high-flying
plane took the equipment above
the troublesome radiation from
water vapour in the Earth's
atmosphere.

Opposite page, below:
A close-up view of the U-2
radiation detector, with the
cover removed. The two large
cones receive radio waves from
different parts of the sky (one

cone is visible in the top
picture, to left of cockpit).
Comparison of the signal strengths
showed that the radio sky is very
slightly 'brighter' in the direction
of the constellation Leo.

to measure the faint radio noise from the outskirts of our Galaxy.

They started, in fact, at a wavelength where they expected no radio signal from the Milky Way, just to test the antenna itself – a large horn-shaped instrument, more sensitive than the usual radio-astronomy 'dishes'. At once, they had a problem. There was a faint residual 'noise' from the antenna; and, try as they could – including the removal of a bird's nest from the horn – they could not eliminate it. Eventually they had to conclude that it was radiation from the sky; and strangely enough, it seemed to originate evenly from all over the sky. If we had 'radio eyes', the background sky would shine dimly but uniformly.

Princeton astronomer Robert Dicke was at that time preparing to look for such a radio background, after performing calculations very similar to Gamow's earlier ones. He quickly confirmed that the radio sky is not black, at a wavelength of a few centimetres. His observations at a different wavelength confirmed that the sky shines like a physicist's 'black body' at a very low temperature. We saw in Chapter 7 that the theoretical black body shines in a way that depends only on its temperature, with the wavelength of peak emission moving to longer wavelengths for cooler bodies. Various later experiments have proved that the whole sky 'shines' as if it were at a temperature of $2 \cdot 7°$ C above absolute zero, or $-270 \cdot 5°C$ ($-455°F$).

Gamow's original prediction at long last returned from the wilderness. There is no other way to explain such a uniformly bright radio sky, unless it is radiation from the beginning of the Universe. We worked out earlier that if we look out in any direction to a distance of 15,000 million light years, we are looking back to the birth of the Universe. The radio background is coming from almost this distance, farther even than the quasars, because it comes from an earlier stage of the Universe's life, before galaxies and quasars were born. At this early time, the Universe was filled with hot gas and radiation, at a temperature of a few thousand degrees, like a present-day star's surface. This radiation peaked in wavelength around the visible and infrared; but the expanding Universe has stretched the waves out until they reach us as centimetre-wavelength radio waves. Penzias and Wilson had unknowingly picked up the radio whispers from the Big Bang in which our Universe began.

Death of the Steady State theory

The discovery of the background radiation turned out to be the death sentence for an alternative theory of the Universe, the 'Steady State'. This popular theory, championed vociferously by British astrophysicist Fred Hoyle, was a serious alternative to the Big Bang theory during the 1950s and early '60s. It was a very appealing theory, because it had

none of the trauma of an initial Big Bang, and no embarrassing questions like 'what happened before the Big Bang?' Many cosmologists mourned its passing, simply because it was so much more elegant than the Big Bang theory. But the Steady State theory came to grief over the background radiation: without a Big Bang, there is no way to produce a uniformly bright background of radio waves.

It's worth devoting a little time to describing the theory, though, because 'Big Bang vs. Steady State' was one of the most important astronomical debates of the century. When the Steady State theory was conceived in 1948, astronomers were still seriously underestimating the distances of galaxies; and, as we've seen, this made the Earth seemingly older than the Universe. But suppose there was no beginning to the Universe. Instead of assuming that the Universe merely looks the same wherever we are, let's also say that it looks the same (on the large scale) whenever we observe it. The Universe has always been the same, and will always be the same. This was the 'Perfect Cosmological Principle', the foundation stone of the Steady State theory.

Since the clusters of galaxies are racing apart, the Universe can only maintain the same appearance if matter appears out of nothing in the space in between the clusters. Even though Fred Hoyle put this part of the theory on a firm theoretical footing, the 'continuous creation' of matter which the Steady State theory needed stuck in the throats of many astronomers, who thought this device no great improvement on the Big Bang which this theory avoided.

Continuous creation did remove the problem of the Earth's great age, though. Hoyle and his colleagues would have said that all the galaxies we see rushing away which, it seems, must once have been coincident with ours in a Big Bang, have actually come into existence from matter created in space comparatively recently. At no time was the average distance between clusters of galaxies less than it is now; nor will the average separation increase with the expanding Universe, for new clusters will appear between the old ones.

Although the Steady State was a beautiful theory, it soon ran into problems with the number of faint radio sources, which the improved techniques of radio astronomers were finding in profusion. Since the Steady State theory said that the Universe has always been the same, the average spacing of radio galaxies should be roughly the same, however far out in space – that is, back in time – we look. Cambridge radio astronomer Martin Ryle found that there were too many faint (distant) radio sources, compared to the bright nearby ones. In other words, radio galaxies must have been closer together, or simply more numerous, in the past. The centrepiece of the whole Big Bang vs. Steady State controversy came to be this question of the radio 'source counts', with little agreement between the two main combatants, Hoyle and Ryle (who were both later knighted).

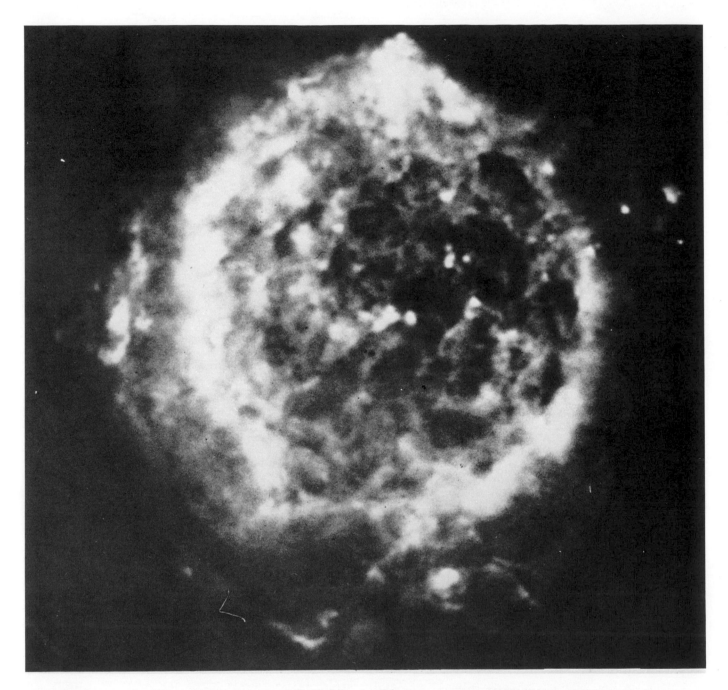

The remains of an exploded star, the radio source Cassiopeia A is shown here in a computer generated 'photograph' which reveals its radio emitting structure. The shell of expanding gas, containing matter from the supernova's core, is more clearly seen than in the optical photograph opposite. Courtesy Mullard Radio Astronomy Observatory.

The background radiation brooked no argument, however, and one by one the Steady State adherents abandoned ship. Some, like Hoyle himself, have ended up with their own theories which combine features of both the earlier rivals; others have accepted the Big Bang theory.

The controversy was not just a waste of scientific man-hours. Cosmologists realized that astronomical observations were reaching the point where rival theories could actually be tested, and choices realistically made. Furthermore, Hoyle and his collaborators have made the first serious study of how heavy elements can be made from hydrogen in stars. The first Big Bang cosmologists, including Gamow, had assumed that all the different elements had been manufactured in the very special conditions of the Big Bang; the Steady Staters had to find an alternative theory. The matter which was continuously created had to be the simplest element, hydrogen, and so all the other elements must be made in stars and spread out into space in supernova explosions. Hoyle, with astronomers Geoffrey and Margaret Burbidge and nuclear physicist William Fowler,

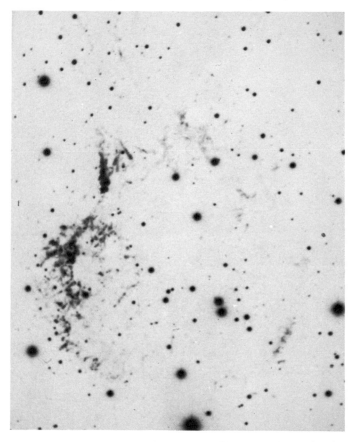

showed how the heavy elements could be produced in the interiors of stars by processes which nuclear physicists have elucidated on the basis of experiments here on Earth. Following this pioneering lead, astronomers have now concluded that all the heavy elements are produced in stars.

But there was one outstanding problem: the second-lightest element, helium, is far too abundant in the Universe. It's impossible to produce as much in stars, while virtually any theory which includes a Big Bang does produce almost exactly the right amount of helium in the Big Bang itself. Astronomers now believe that the hydrogen and almost all the helium in the Universe are relics from the Big Bang, while the heavier elements were made in stars. Hoyle's calculations on heavy-element 'building' in stars have stood the test of time, even though the Steady State theory itself is apparently headed for oblivion.

The Big Bang
The evidence for a Big Bang is now overwhelming. And with the available evidence on the age of the Universe, the helium abundance and the temperature of the background radiation, cosmologists can work back in time towards the moment of the Big Bang itself. Before the galaxies formed, the Universe was filled with a more or less uniform 'gas' of particles and radiation, and the physics can be worked out quite straightforwardly.

The actual moment of the Big Bang itself – time equals zero exactly – is shrouded in mystery. (We'll talk only in terms of an open Universe in what follows, but the essential results are the same for a closed Universe as well.) Although the Universe is always infinite in size, the farther back in time we go, the more matter is packed into a particular volume (say a matchbox volume); and the hotter is the surrounding radiation. We can see instinctively – and Stephen Hawking and others have proved it – that at the moment of the Big Bang, there was literally an infinite amount of matter packed into each matchbox volume. And the background radiation must have had an infinitely high temperature.

Did this really happen in our Universe? No one knows. Hawking's calculations are based on Einstein's gravitational theory, which describes gravitation in a smoothed-out way. For the gravitational fields we usually meet in the Universe today, it's an excellent theory. Theorists like Hawking always bear in mind, though, that quantum theory insists that gravity is transmitted by virtual particles, the gravitons. When matter is highly compressed, the gravitational fields become so strong that Einstein's equations must break down, and physicists have to think in terms of gravitons. Unfortunately, a proper quantum theory of gravity still eludes the theorists, despite decades of dedicated labour; and so physicists are uncertain about the density of extremely high-density matter. We met this problem in Chapter 9, in the controversy over the 'singularities' in the centres of black holes. Until we have a better theory of gravitation at high densities, cosmologists can't be certain that the Big Bang actually did start with an infinite density, and infinitely high temperature.

Without this improved theory, we certainly can't see back to the Big Bang itself, to 'time equals zero'. But we can get a glimpse of what happened only 1/10,000,000,000,-000,000,000,000,000,000,000,000,000,000th of a second after the Big Bang! The temperature and the density were unimaginably high, but they were not infinite. Theorists can calculate both these quantities and set down definite, enormously large figures: a temperature of 100,000,000,-000,000,000,000,000,000,000° C and a density of 1,000,000,000,000,000,000,000,000,000,000,000,000-000,000,000,000,000,000,000,000,000,000,000,000,-000,000,000 times the density of water. And cosmologists can calculate how the Universe cooled, and the density fell, as space expanded from that moment onwards.

What sort of particles filled the Universe then is not so certain, though. Our present theories of the different particles and their interaction is only proved for a certain range of densities – remember, astrophysicists run into trouble when trying to predict what happens in the very core of a 187 neutron star, where matter is packed to a density 'only' 1,000 million million times that of water. What happens at still higher densities depends very much on how the fundamental particles behave when the amount of energy around is extremely large. Experiments being carried out in the high-energy accelerators will provide some of the answers.

With this concentration of energy in the early Universe, pairs of particles and antiparticles can spring into existence with no trouble: the radiation energy can readily appear as the equivalent amount of mass. To find out what particles were present, physicists will have to learn more about the fundamental particles, the quarks and leptons themselves. Are there only a handful of each; or is there a whole succession, with increasingly high masses? These possible very heavy quarks and leptons (with their antiparticles), too massive to produce in the comparatively puny energy source of our terrestrial accelerators, would have been almost as common as the lighter, everyday ones in the early Universe.

The role of the forces is obscure, too. After the instant we detailed above, gravitation settles down to its familiar behaviour according to the Einstein equations. But the other three forces should be revealing their basic similarities, if present-day theories linking them are correct. Before the temperature cools to 3,000 million million degrees, the weak and electromagnetic forces will be behaving in exactly the same way, according to the Weinberg-Salam theory which links them. As the Universe cooled, the forces 'froze' out, and the electromagnetic force ended up a thousand times stronger than the weak force.

Steven Weinberg compares this separation to water freezing, in an analogy which is not exact but gives a flavour of the theoretical reasons. We could put it this way. Water vapour in the atmosphere is spread out pretty evenly, and a sheet of glass left out on a cool night acquires a uniform coating of dew deposited from the air. But if the glass is cold, the ice which forms on it grows in beautiful swirls and curls: when the water vapour condenses out, a pattern forms which didn't exist before. In a similar vein, the weak and electromagnetic forces could have been exactly the same in the early Universe, but 'froze' out in the relatively simple 'pattern' that one ended up a weak, short-range force and the other a stronger, long-range force.

The strong force between quarks poses another problem. We saw in Chapter 6 that when quarks are close together the strong force effectively 'switches off'. In the early Universe, all the quarks were close enough together for the strong force to affect them very little. But at some stage, the density dropped too low, and the strong force began to marshal them into heavier particles, particularly protons and neutrons (and antiprotons and antineutrons) Somehow

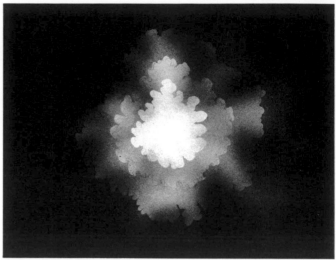
Big Bang (time=0)

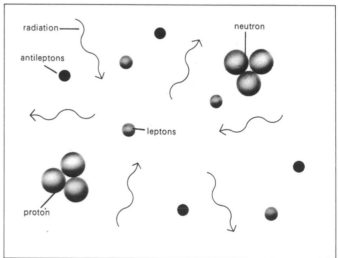
time=1/100 second

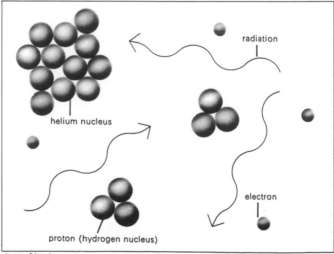
time=3½ minutes

188

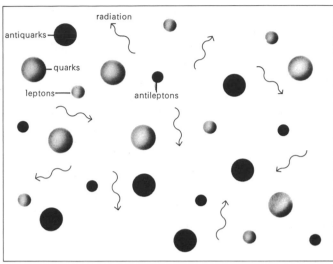

a moment after (see text)

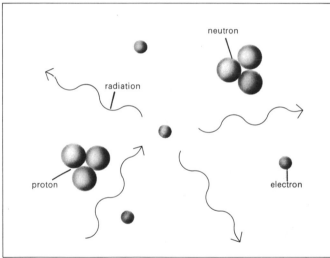

time = 2 seconds

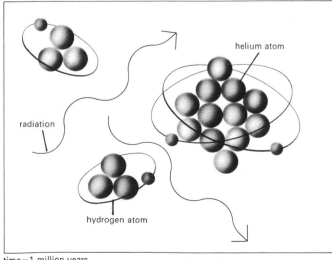

time = 1 million years

it was so efficient that no free quarks were left over. Sensitive experiments to look for loose quarks have been singularly unsuccessful, despite a false alarm in 1977.

Cosmologists must await the results from particle accelerators, and the corresponding refinements in fundamental theory, before they know accurately what happened in this maelstrom of the early Universe. If looking out in space is looking back in time, surely we could use distance as a sort of time machine? To some extent this is true: if we could detect the background of neutrinos which should survive from this period, and the probable background of gravitons, we could learn a lot about the first few moments of our Universe. Unfortunately, our present techniques are nowhere near sensitive enough – it's difficult just to pick up neutrinos from the Sun.

And the radio background tells us very little. The early, hot Universe was in a state of heat balance. Radiation and particles were so densely packed that they interacted continuously, and everything was effectively at the same temperature. It was rather like the 'black body' which emits radiation appropriate to its temperature, but in this case the concentrated radiation controlled the temperature of the matter. Because of the heat balance, all the details of what happened in the first few minutes have become smeared out. Steven Weinberg aptly puts it this way: 'It is as if a dinner were prepared with great care – the freshest ingredients, the most carefully chosen spices, the finest wines – and then thrown all together in a great pot to boil for a few hours.' The background radiation is a bland amalgam of the exotic physics that must have been going on.

The first one-hundredth second onwards

After the first one-hundredth of a second, however, the Universe had cooled to a 'mere' 100,000 million degrees, and present-day theories can tell us what must have happened. By this time, antiprotons and antineutrons had disappeared from the Universe. Collisions with protons and neutrons had all along been causing mutual annihilations, but when the temperature and density were high, new pairs appeared spontaneously. Now it's too cool for this to happen and the remaining antimatter particles disappeared for good (except for positrons, of which more later).

But if all the protons and antiprotons were made in equal numbers as 'pairs', then all the protons should have disappeared, too, to leave pure radiation. We all know that there are protons (and neutrons) around, making up stars, planets and our own bodies; and the reason why Nature seemed to prefer matter to antimatter is still one of the unsolved riddles of cosmology. Her favouritism wasn't very marked, in fact: in the beginning there was roughly one extra proton for every thousand million proton-anti-proton pairs; but when these pairs were annihilated, the odd pro-

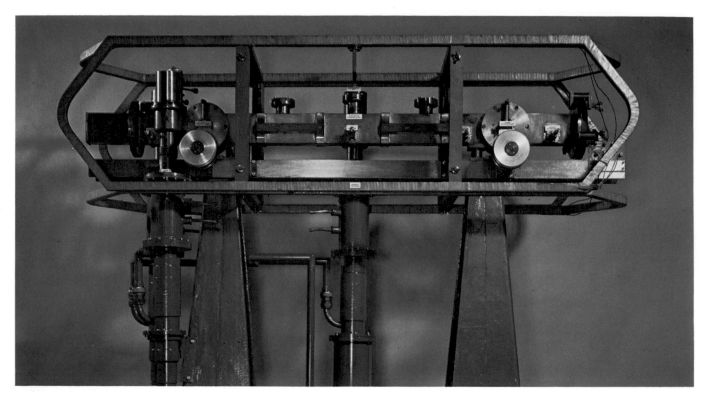

tons were left to make up our Universe. The same argument goes for neutrons and antineutrons, too. As time went by, though, the excess neutrons gradually decayed to protons (and electrons and antineutrinos), just because the neutron is not basically a stable particle. Its average life of about fifteen minutes is a lot longer than the age of the Universe at the time we're talking about.

At this early time, between one-hundredth of a second and a few seconds after the Big Bang, the Universe was filled mainly with radiation, neutrinos and electron-positron pairs. The temperature and density were still high enough for pairs of these very low mass matter-antimatter pairs to spring into existence. Scattered around, and very unimportant now, were the relatively small number of protons and neutrons.

At around two seconds, the Universe had cooled to the point where electron-positron pairs cannot spontaneously appear. They now just annihilate each other, leaving a small residue of electrons, to balance the positively charged protons. Radiation now dominates the Universe. There are around a thousand million electromagnetic photons to every proton, neutron or electron; and these energetic radiation particles have the whip hand, controlling the temperature of the Universe as it cools over the following million years.

What happens to the matter particles during the first few minutes of this 'radiation-dominated era' is important to us, because it's when protons and neutrons came together to make helium nuclei. It was a step process: first a proton and neutron must join to make a nucleus of heavy hydrogen, deuterium, and then another proton and another neutron are added (in either order) to make the two-proton, two-neutron nucleus of helium. Deuterium is a feebly bound nucleus, and in the chaos of the Universe's early stages it would have been broken apart as soon as it formed. But when the Universe is around three and a half minutes old, the deuterium can survive long enough to capture the extra particles, and end up as the extremely stable helium nucleus. Practically all the neutrons still around now team up with protons, and end up in helium. Knowing what fraction of neutrons have survived this long, it's quite easy to work out that around a quarter of the matter in the Universe will end up as helium; and, as we've seen, this is just the amount of helium which the oldest stars contain.

There are two more important dates in the Universe's life. Going on a long way, from three and a half minutes to its millionth birthday, the Universe has now cooled down to only 3000°C (5400°F), cooler than the Sun's surface. Now electrons can combine with the nuclei around, to make atoms of hydrogen and helium, without being immediately kicked away again by the now much weaker photons. Unlike the previous plasma of free nuclei and electrons, the gas is now transparent to radiation. Although the radiation remains around for the rest of the Universe's history, it is always weakening, and never again has a serious effect on the matter. When we look out into space and back in time, the farthest our instruments can peer is back to the Universe's millionth birthday. Farther away than that (farther back in time) the Universe was filled with opaque gas, or more accurately, plasma. The radiation filling the Universe when it was a million years old has spread out through the now transparent Universe, and as it has spread it has faded. This is the faint whisper first picked up by the big satellite antenna at Holmdel in 1965.

The last important date in this story is when the gas began to clump together to make galaxies. Astronomers are not sure exactly when or how this important event occurred. When the Universe was a few hundred million years old, individual clouds of gas began collapsing as protogalaxies, and the gas within them began to form into stars. Heavier elements were synthesized, explosions threw them out into space, and later stars could form with planetary systems. Our attention moves from the whole Universe – now an expanding collection of clusters of galaxies – to the galaxies themselves and their contents: astronomy takes over from cosmology.

The end of time

Our expanding Universe began in a Big Bang. We have called the moment of the Big Bang 'the beginning of time'; we have timed the Universe's various birth pangs from that instant. You might think this is a very arbitrary approach: surely there must have been time before the Big Bang. As we'll see later, some cosmologists think the Universe is 'oscillating': expanding and contracting in long cycles, so that our Big Bang was just the occasion when the Universe last reached its minimum size. Before that it had contracted, and even further back we would find it expanding from a previous Big Bang. The Universe would have existed since time immemorial, even though all the particles we have around us today, and perhaps even the present laws of Nature, have only existed since the last Big Bang.

Some find this oscillating Universe, with no beginning to time, a philosophical comfort. But is it so strange to think that time only began 15,000 million years ago? After all, there is an absolute zero of temperature: we can never freeze anything below $-273 \cdot 15°C$ ($-459 \cdot 7°F$). Perhaps there is a limit on time, too, at 15,000,000,000 BC. Our everyday experience runs contrary to this: time is an 'ever-rolling stream', some great power which moves relentlessly on, from the infinite past to the indefinite future. Yet everyday experience is not always a reliable guide. The floor under your feet seems solid; yet it consists mainly of the empty space between the electrons and nuclei of atoms. Outside the everyday scale, common sense cannot always be trusted.

Scientists approach the Universe warily these days. To Newton there was absolute space and time: Einstein taught us that space and time are changed by rapid motions and by the gravitational pull of massive bodies. To define a scientific concept, scientists talk in terms of the way we measure it: if we cannot measure, or even detect something, we are on thin ice in saying that it exists. This is the case with time: at present we can define time in terms of heartbeats, or of our planet's spin, or more fundamentally by the vibrations of atoms. The vibration rate of the caesium atom is now used to define our time scale, and standards laboratories around the world follow this atomic time: when our gradually slowing Earth gets out of step, they add a 'leap second' to our civil time.

'Before' the Big Bang, there were no heartbeats, no spinning planets, and not even any atoms. Even if our Universe had another oscillating phase then, there's no way that we can look back beyond the Big Bang to see if atoms identical to ours existed. There is no logical way to define time before the Big Bang. And so there's no logical fallacy in saying that time itself began with the Big Bang, along with space which our Universe occupies, and the particles of matter within it.

What of the end of the Universe? Will there be an end to time? Cosmologists are divided when it comes to discussing the Universe's far future. Although the Universe is expanding furiously now, there is a weak gravitational force attracting each receding cluster of galaxies to every other. The fate of the Universe depends on whether this gravitational tension will be enough to slow the speeding galaxies; halt them; and then pull them back together again in a rapidly shrinking Universe. If astronomers knew exactly how much matter there is in the Universe now, the cosmologists could easily decide whether there is enough to stop the expansion, or if the galaxies will speed apart forever. The culprit is the 'missing mass'. If the galaxies we see are the only matter in the Universe, their gravitational attraction is far too small to stop the Universe expanding. Yet it's possible that a lot of invisible matter lurks around and between the galaxies, as we saw in the last chapter. Astronomers are hard at work following up alternative approaches to working out how much invisible matter there is; but, so far, there's no general agreement.

If the Universe is slowing down in its expansion, then, when we look out to great distances and early times, we should see the galaxies there (and then) moving apart faster than they are now. Hubble himself started looking for this effect, and Allan Sandage has pursued it to the limits of present-day techniques. But the results are disappointing. He can only say that the Universe will continue expanding for at least another 50,000 million years, but the observations cannot help us to decide whether it will then begin to contract.

Other astronomers have tried a plethora of different methods. Simply extrapolating the missing mass of clusters of galaxies up to larger scales is one approach. Another way is to use the focusing effect of the 'bent' space in Einstein's models for the Universe. We saw earlier that there are two kinds of space, 'open' and 'closed'; and, for the simplest equations of cosmology, these correspond exactly with a Universe which is continuously expanding, and one which will eventually start to contract. The Universe which expands forever is the open model, which is always infinite in 'size'. On the other hand, the Universe which has enough matter to slow the expansion to a standstill is the 'closed' model which is bent around on itself – where, if it weren't for the complication of expansion, you could set out in a straight line through space and arrive at your own back door.

In both kinds of Universe, rays of light are focused slightly as they travel through the empty space between the galaxies. If you moved away a standard-sized object – for example, a ruler – you would first see it appear to shrink, as you'd expect. But once it passed a distance of roughly 10,000 million light years, you'd see it begin to appear larger again, even though it was always getting farther away! This focusing effect is, not surprisingly, stronger for the 'bent', closed, Universe, and astronomers have looked at various objects – clusters of galaxies and radio galaxies in particular – to try to see how their size depends on distance. The answers aren't conclusive, though, because Nature hasn't provided standard, identical rulers for us to measure.

. The most intriguing method of predicting the end of the Universe depends on something that happened very near its beginning: the production of deuterium. Almost all of these heavy hydrogen nuclei picked up extra protons and neutrons to turn into helium. But a tiny proportion remained, and the theory relates this fraction to the density of matter at the time. Ultra-violet satellites have detected the spectral lines of deuterium gas in space, and their strength seems to show the Universe to be comparatively low in density: cosmological theory relates the present amount of deuterium to the density of the early Universe, and from this it's possible to work out that the Universe will expand forever.

The whole question is far from settled yet, though all the different methods should eventually yield the same answer. For the moment, let's look briefly at both possible outcomes.

In a closed Universe, the expansion is gradually slowing down; and, some time in the future, say around 75,000 million years from now, gravitation will win outright. The Universe will start to collapse again. Clusters of galaxies will begin to move together, faster and faster, until 160,000 million years from now the entire Universe will fall together in the 'Big Crunch'. Matter will be squeezed tighter and tighter, and as it does so, the gravitational force will grow increasingly strong, and theory predicts a singularity of infinite density. Some 'Big Crunch' cosmologists believe that the Big Crunch is another Big Bang; the Universe will immediately begin to expand again on another cycle of existence. This new Universe could have different kinds of matter, different forces and even different laws of physics from ours.

It's probably fair to say, however, that most astronomers now think that the Universe is comparatively empty: it is open and will expand forever. In this case, too, the prospect is bleak for our descendants. Although they won't have to face the impending crunch helplessly, they will find themselves in the trap of the 'heat death' which Lord Kelvin foresaw in 1851, before other galaxies and the expanding Universe were known.

It will be a protracted death. To describe it in terms of years would be inconvenient; instead, let's call the age of the Universe so far – all 15,000 million years of it – one 'cosmic second'. On this new scale, we now exist at a time of one cosmic second after the Big Bang, and our Sun and Earth are about one-third of a second old.

Within each galaxy, stars are dying all the time. New stars are born from gas in spiral and irregular galaxies, and this uses up the available gas supply. Although dying stars puff off or explode some gas into the space between the stars to replenish the gas, the amount they return is always less than they originally took. The balance ends up as the compact star corpses, white dwarfs, neutron stars or black holes. So the gas supply is gradually diminished, and star birth will eventually cease. Galaxies will become star cemeteries, with the dark, dead star remains still revolving in their orbits.

After a few cosmic minutes (say, ten thousand million million years of ordinary time), our Sun has long since faded out, as a cooled-off, black 'white dwarf', circled by a frozen retinue of planets. Even if our descendants have moved on to the planets of younger stars, these stars, too, will have gone a few more cosmic minutes from now. Our Galaxy will be completely black and cold, with no source of light or heat to keep life's processes running. It's difficult to see how life could survive when Nature's sources of energy have given out.

But let's not be too pessimistic. Climbing into the time machine of cosmological theory, we find that galaxies of dead stars are not the end of the story. As they orbit, the stars are losing energy all the time, as gravitational radiation (the gravitational version of the electromagnetic radiation which is produced when an electric charge accelerates). All orbiting bodies – the Moon, the Earth and the Sun, all in very different orbits, but similar in this respect – are losing energy in this way right now, but at an incredibly slow rate. Over ordinary spans of time, even over the age of the

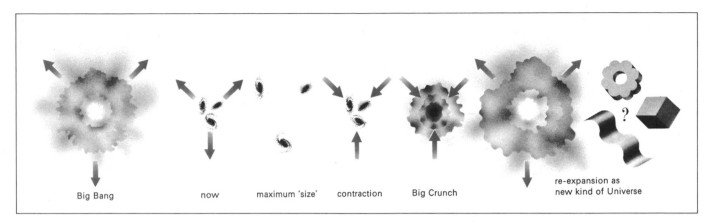

Big Bang   now   maximum 'size'   contraction   Big Crunch   re-expansion as new kind of Universe

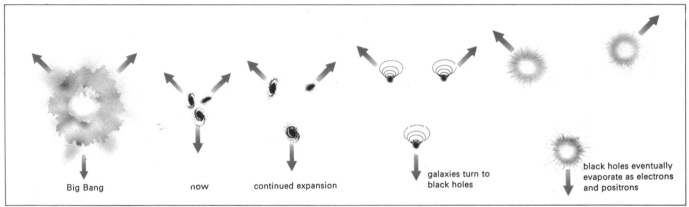

Big Bang   now   continued expansion   galaxies turn to black holes   black holes eventually evaporate as electrons and positrons

Universe, one cosmic second, it doesn't make a measurable difference to the orbit.

But when we start talking about time spans of *cosmic* years, the orbits of the stars around the centre of the galaxies *are* gradually shrinking. If we wait around for a very long time – and that means a million million *cosmic* years – we shall find that the dead stars in our Galaxy (including the ember that was the Sun) have spiralled right into the centre, to make a massive black hole.

In the clusters of galaxies, the individual galaxies will spiral in towards the cluster centre, and fall together as an even heavier supergalactic black hole. The Universe will consist of huge black holes, each as heavy as galaxies or clusters of galaxies, and now separated by enormous tracts of empty space.

There is a final act in the Universe's story. Remember that Stephen Hawking showed that black holes slowly evaporate, effectively losing their mass as an outflow of electrons and positrons, slowly at first, but building up to a final explosion. The heavier a black hole, the longer the evaporation takes. A star-mass black hole would last over a period of time much longer than we have considered so far, and these black holes will disappear intact into the galaxies' central black holes. But these galactic and supergalactic black holes are evaporating, although even more slowly.

These holes gradually turn into electrons and positrons over a period of some 1,000,000,000,000,000,000,000,000,-000,000,000,000,000,000,000,000,000,000,000,000,000,-000,000,000,000,000,000,000 *cosmic* years!

At this mind-boggling date in the future the crystal ball begins to go hazy. Our time machine shows us that the Universe contains almost only electrons and positrons, separated by vast distances. In the still expanding Universe, it's unlikely that these particles will ever meet up and annihilate. And with them, spread out unbelievably thinly, will be the photons of the radio background, now cooled down and stretched to wavelengths longer than we can possibly imagine.

These photons perhaps have the final word. Ever since matter took it upon itself to form atoms, and pursued its own course, independent of radiation, only a million (Earth) years after the Big Bang, the photons of radiation have been mainly passive onlookers. From galaxy and star formation, and the era of life – for that's all it can be – to the death of matter in black holes, and its resurrection as scattered electrons and positrons, these photons have remained aloof. The stoical background radiation has missed out on the triumphs and tragedies of the Universe. But some of these photons are kindly falling upon our radio telescopes, and telling us how our Universe began.

The frontiers of science are always expanding. On the first Moon-landing, in 1969, Edwin Aldrin deployed an experiment to sample the solar wind of charged particles. Since then, our knowledge of the inner solar system has grown enormously; unmanned probes should soon bare the secrets of the outer planets. And space is but one of many frontiers to be explored.

12: The Continuing Quest

The entire realm of the Universe is the research field of modern scientists. We've seen how experimenters and theorists are collaborating over the whole range of size, age and energy that the Universe places in front of us. And the strides in our understanding have been colossal.

The legion of scientists working now outnumbers all previous generations of scientists combined. Add to this the multi-million-dollar particle accelerators and telescopes which are pushing out the frontiers of knowledge, and also computers, which remove the tedious burden of long calculations; and the speed of today's developments may not seem so surprising. The traditional picture of an isolated scientist stumbling across a breakthrough in his own laboratory is certainly no longer true. Large teams of scientists are involved in most research projects, each member contributing his special skills and assisting the others to speed analysis of the results.

Despite a certain amount of healthy rivalry for new findings, modern communications have produced a remarkable degree of international co-operation. There is a genuine feeling of an international community, particularly in a science like astronomy, where there is only a relatively small number of active researchers. It's refreshing to experience the spirit of total openness which characterizes international scientific conferences.

One danger of this research impetus is that the results may not percolate down to the general public who are ultimately paying for this exploration of the unknown. Researchers in the front line naturally concentrate on their research projects, and – with a few prominent exceptions – they have little time to spare for popularizing their work. And since the realms of quarks and quasars are far from our 'common-sense' world, these results can't just be summarized in news flashes: they need explanation before the average person can understand their importance.

Even if the uncommon-sense of the Universe makes for difficulty in communicating ideas, it is fascinating in itself. We are being far too parochial if we think it should be easy to understand the very small, the very large and the very powerful in terms of things we are used to. Our everyday world of experience is just one level of the Universe. If, for example, we usually lived on the small scale, where the Uncertainty Principle would be paramount, and we were transported to the human world, it would seem nothing short of miraculous that a golfer could accurately predict where his ball will go!

The essence of scientific advance is to produce theories which agree with past experimental results, and which will also predict correctly what will happen in future experiments. If these predictions are verified, the theory must be a fairly accurate description of what is going on. Scientific theories must predict actual numbers to compare with

experiments; and this, of necessity, makes them mathematical. To other scientists working in the field, this is no inconvenience; but when it comes to explaining what a theory in some uncommon-sense part of science means in simple language, there are bound to be difficulties.

The theory of relativity is perhaps the most famous example. Largely because it is couched in mathematics, it gained a reputation for incomprehensibility; and even when explained in simple terms – perhaps best by Einstein himself – its apparent contradictions with common sense confused many readers. Quantum theory is another good case, for its equations predict that a parcel of energy will sometimes behave like a 'wave' and in other situations as a 'solid particle'. The wave-particle contradiction doesn't exist in Nature itself, but only crops up when we try to explain this exotic situation in common-sense terms. A theory must be logically constructed – and this is where the mathematics comes in – so that it doesn't contradict itself in any way, and also fits in with the confirmed framework of established theories; otherwise, its conclusions would be worthless.

The power of the way that science approaches the Universe has been evident throughout this book. The whole Universe can be explained in terms of a few types of particles; a few forces; and a few laws of Nature. From the small world of quarks to the enormous powerhouses of the quasars; from a fraction of a second after the birth of the Universe to today's complex community of galaxies with their constituent stars, gas clouds and planets, everything seems to be related.

On the small scale of quarks, Nature seems relatively simple; and again, when cosmologists think of the Universe as a whole, they regard galaxies as merely evenly distributed lumps of matter in a simply expanding Universe of empty space. The very uniform distribution of the background

An aurora results from the Sun's
outward streaming 'wind' being
channelled by the Earth's
magnetic field into the atmosphere.
But as in most scientific research,

detailed investigations have shown
the situation to be more
complicated: further research is
needed before we fully understand
the Northern Lights.

radiation from the early reaches of the Universe shows that on the large scale, our Universe is extremely even – indeed, astronomers find it hard to work out why the originally very uniform gas filling the Universe clumped together as proto-galaxies. And the stars, too, despite their impressive appearances, are pretty simple things. The height of complexity in the Universe seems to be reached by living creatures, where intricate chemical reactions and complex molecules have produced the phenomenon of life.

Are quarks the end?

Despite all this harmony, modern scientists have by no means exhausted the treasury of Nature's secrets. If they had, there would be no point in making further observations of the Universe. There will always be new facts to fit in; new experiments which will conflict with established theory. Present-day science is very much at the point where the framework of the Universe seems to be established; no one could claim that the details are finalized.

In the world of particle physics, we must wonder about the apparently 'superfluous' particles. To construct the Universe, we only need the 'up' and 'down' quarks, and the electron and its neutrino. Yet there are at least two other quarks, and two other leptons. These may be connected in some subtle way with the manner in which Nature displays four apparently different forces, which are probably merely different manifestations of just one basic force.

There are four well-known quarks, with another two probables. But there's an added complication we haven't mentioned so far. The quarks must have a new kind of 'charge', in addition to their electric charges. Quantum laws prohibit two quarks in the same particle from being identical; yet in a particle like the omega minus there are three apparently identical 'strange' quarks. Theorists believe – and there are observations to support the idea – that the quark 'charges' come in three varieties (unlike the electric charge, which has two kinds, positive and negative). In a particle like the proton or the omega minus, each quark has a different one of these three 'charges'; and so, for the purposes of the quantum theory, even the three 'strange'

197

quarks making up omega minus are not identical. (The 'charges' are usually called 'colours', to distinguish them from ordinary electric charges.)

Immediately we find our quark family of four (or six) has to be expanded. Since each type of quark can adopt one of three colours, the family is extended to twelve (or eighteen) members. This is becoming rather a large number of 'fundamental' particles. Are the quarks themselves simply reflecting some deeper unity, still to be unearthed? And are there a definite number of quarks; or, as we build accelerators operating at higher and higher energies, will we keep on finding more and more different kinds? Particle physicists are convinced that there must be a pattern of some kind behind it all.

We can ask the same questions about the leptons, too. Perhaps new, heavier leptons will establish what the apparent link between quarks and leptons means: why there's a symmetry of 'up' and 'down' quarks with electrons and their neutrinos; and between 'strange' and 'charmed' quarks with muons and muon neutrinos.

All these questions should be bound up with the problem of unifying the forces – certainly this is the theorists' hope. The weak and electromagnetic forces are successfully tied together by Steven Weinberg and Abdus Salam's theory; although, in this book, we've generally mentioned the two forces separately, because their effects are so different. This unified theory was vindicated in 1978 by a crucial experiment carried out at the big Stanford electron accelerator.

This experiment relied on an odd quirk of the weak force,

discovered back in the 1950s: it is left-handed. A neutrino produced by a weak interaction has a 'spin'; so, as it's flying through space, it's performing a sort of corkscrew motion. We all know that a corkscrew has a 'handedness': an ordinary (right-handed) screw goes in if you turn it clockwise; but it's equally possible to make a left-handed screw which comes out when it's turned clockwise. Screw extractors, used by people who repair machinery, work on this principle. A neutrino always performs a left-handed screw motion, while its antiparticle, the antineutrino, travels with a right-handed screw turn.

If the electromagnetic and weak forces are really related, the argument went, then there should be a very slight 'handedness' effect in electromagnetic interactions, too – theoretically, even in electromagnets, or a rubbed comb picking up pieces of paper, although we couldn't possibly detect it in such circumstances. Experimenters at Stanford shot electrons at protons, and discovered that this simple electromagnetic experiment – the electric force of the positive proton attracts the negative electron – has a slight 'hand' preference. The emerging electrons are not quite evenly distributed between right- and left-handed spins; and the discrepancy was exactly the amount that Weinberg and Salam had predicted. From June 1978, there was to be one fewer force of Nature to consider as fundamental.

But many problems are still outstanding; and, over the next few years, particle physicists will be cracking down on the relationships between the forces and the particles, aided by the new generations of powerful accelerators now

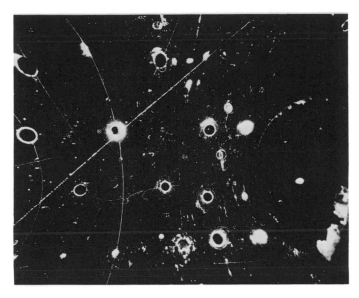

scientists have thought of electricity and magnetism as fundamentally the same (electromagnetic) force for over a century. To make a very simple analogy, a cylinder looks different when we look at it from different directions: end-on, it's a circle, while seen from the side it's a rectangle. It's not really surprising that the basic force of Nature can behave in different ways when it is involved with the various particles in different circumstances. It appears to the quarks as the tightly binding strong force, to interactions involving the leptons as the decay-controlling weak force, and to all situations involving electric charges as the multi-faceted electromagnetic force of electricity, magnetism and light.

Toward 'ultimate reality'

Present-day gauge theories tell us something of how the Universe is made up at an even deeper level, where we can view the Universe as a complex mixture of 'complete order' and 'complete disorder'. Again, we can't understand this very well in common-sense terms, but we can draw on another analogy. A sugar cube is highly symmetrical: its six-sided shape is beautifully ordered, and there's no logical way to prefer any one face to any other. Turn the cube around, and all the faces are identical. But if we drop it on the table, one face must finish upwards. The symmetry is 'broken', for there's now a distinguishing feature (only one face gathers dust!). A degree of disorder has entered our ordered sugar-cube theory – not complete disorder, but a contribution which, we could consider, makes the cube more interesting.

In the real world of fundamental particles, complete 'order' would mean that the force particles all have zero mass. This is certainly true for the photons of the electromagnetic force, and the gravitons, but the weak-force W and Z particles do have quite a large mass, probably some 150,000 times heavier than the electron. Perhaps, surprisingly, the 'disorder' of the Universe can make particles massive. In very rough terms, we can think of the W and Z particles as being photons which have 'acquired' mass because a certain amount of disorder has crept in. This is the essence of the Weinberg-Salam theory: the basic particles which carry the weak and electromagnetic forces are the same, but the former involve some disorder which gives them mass, and therefore (as we saw in Chapter 6) means that they only act over a short distance.

Edinburgh physicist Peter Higgs first combined this subtle kind of disorder ('spontaneous symmetry breaking') with gauge theories in 1964, and set the scene for present-day theories. His calculations show that the 'disorder' may actually appear as a new, relatively heavy kind of particle, the Higgs meson, whose behaviour would be radically different from the quarks, leptons and 'force particles' now

being designed and built. The first priority is to find the W and Z particles, which carry the weak force. The gluons, which exchange the far more powerful strong force, are on the agenda too. Theory predicts that there are eight types of gluon, a number related to the three different kinds of strong 'charge' or 'colour' which quarks can have.

Despite the seeming proliferation of 'fundamental particles' and 'particles which carry forces', there is now a basic framework in particle physics. The 'building block' particles, the quarks and leptons, spin on their axes with exactly half a natural unit of spin. The particles which carry forces (and radiations) – the photon, the W and Z particles, the gluons and the gravitons – have an exact whole number of spin units: two for the graviton, and one for the others. Physicists can make sense of this seemingly strange fact; and theories depending on it are known as 'gauge theories'. Weinberg and Salam successfully used a gauge theory to unite the weak and electromagnetic forces; and another gauge theory describes how the gluon strong force holds the quarks together in a proton or other baryon. Physicists expect that there is a more general gauge theory which will encompass both of these, and so will combine the strong force with the weak and the electromagnetic. (Gauge theories were revived by a young Belgian physicist, then a doctoral student, called Gerard 't Hooft. The problem was theories involving massive 'spin one' particles were not renormalizable: that is, they did not make mathematical sense, because they gave infinite results. He showed that a particular class of theories involving massive particles gave finite results, and developed new techniques to deal with them.)

Three of Nature's four forces will then be housed under the same roof. We will be able to regard them merely as different manifestations of one very basic force, just as

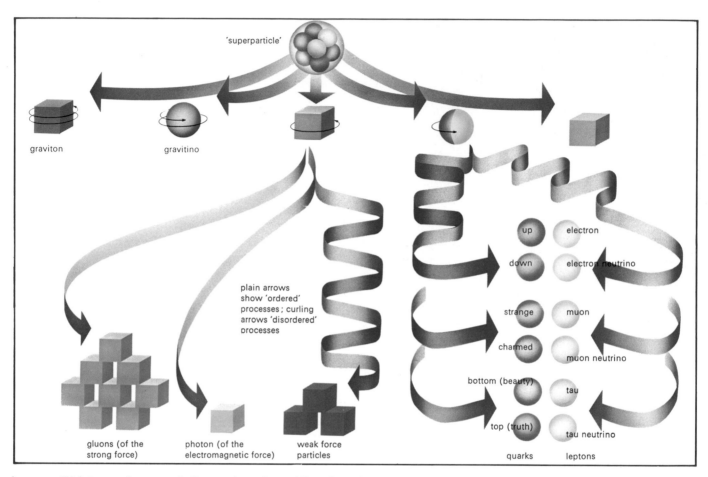

'superparticle'

graviton

gravitino

plain arrows show 'ordered' processes; curling arrows 'disordered' processes

gluons (of the strong force)

photon (of the electromagnetic force)

weak force particles

up — electron
down — electron neutrino
strange — muon
charmed — muon neutrino
bottom (beauty) — tau
top (truth) — tau neutrino

quarks — leptons

known. Weinberg, for one, believes that these 'disorder particles' may turn up in particle-physics experiments within the next decade; although some theorists believe they are just a mathematical figment of the theory.

In the field of particle physics, we are seeing a shift away from talking of 'what things are made of'. It seems that quarks and leptons are the basic building bricks. The relationship between these particles and the forces is becoming top priority now; and physicists hope that we may reach a deeper understanding of how 'particles' and 'forces' may be related to still more basic concepts, such as the conflict between 'order' and 'disorder' ('symmetry' and 'symmetry breaking' in the physicists' language). The type of 'order' which physicists refer to here is closely linked to the 'laws of Nature', like the law which says that electric charge cannot be created nor destroyed. So in effect, physicists are trying to extend the laws of Nature we know, and anticipate that by adding an extra 'disorder' factor, it will eventually be possible to prove quite naturally that the forces and particles of the Universe should be just the ones we find. And such deeper theories should yield answers to questions on which present-day theories stay mute: how many quarks and leptons are there? And why do the individual quarks and leptons have the particular masses that we measure?

Whither gravitation?

We have ventured on to speculative ground, following some of the latest developments in particle physics. But there is one force these scientists cannot measure in their experiments: the incredibly feeble force of gravitation. Theories of gravitation will undoubtedly change in the next few years, for, as we've seen several times in this book, there is as yet no way of combining Einstein's theory of general relativity with the quantum theory of gravitons. In astronomy, general relativity works where Newton's original theory and many other contenders have failed. But, on a basic level, graviton exchange must be involved. And in the strange conditions of a 'singularity' within a black hole, or the Big Bang at the start of the Universe, general relativity must break down. In such circumstances, only a fully-fledged quantum theory of gravitation will suffice.

No one yet has much idea how this theory will eventually emerge, or what strange predictions it will make. When it comes, physicists will have to buckle down to the task of combining the gravitational theory with that of the other

The beautiful spiral pattern in the discs of galaxies is caused by gravitation, the least powerful force in nature, and the one whose basic nature is still not understood. This photograph of M83, from the Anglo-Australian Observatory, has been specially printed to show details in the spiral arms – compare with the normal print of M83 on p 164.

three forces. This is the pursuit of a 'unified field theory' over which Einstein spent many fruitless years (although he laboured under the burden of his era's comparative ignorance of the other forces).

One strong contender, which would achieve both a quantum gravitational theory and a union with the other forces – and, into the bargain, link these firmly with the quarks and leptons – is a recent development called *supergravity*. It takes the idea of a fundamental 'order', and extends it to cover both 'particles' and 'force particles', as a supersymmetry: lo and behold, gravitation appears both as a 'force particle' and as the kind of field which general relativity describes. Supergravity is an elegant way of combining forces, particles and the basic idea of 'order' in the Universe; but it's only in its early stages.

Restless space

Some physicists, led by American John Wheeler and Russian Andrei Sakharov, want to start at the most fundamental level possible, and build upwards, in the hope of arriving at our known forces and particles from the empty space which is the background to everything which happens in our Universe. They point out a simple analogy in the everyday world. A rubber band is 'springy' because of the way its constituent molecules are arranged; yet, however accurately we measure the band's springiness, we can never deduce from this that rubber bands are made up of molecules. Only when we know already that matter is made of molecules (from countless experiments in chemistry and physics) can we show how these molecules give rubber its strange property.

Perhaps gravitation is some sort of 'springiness' of space, due to an unknown small-scale structure of space itself? On this view, however hard we study gravitation, we can't learn anything more basic about it. Physicists must start by studying what space *is*. But surely, we might argue, space is empty space is empty space! Not so, according to modern physics. Remember we discussed in Chapter 6 how forces are conveyed by 'virtual' particles, force particles which appear and disappear in such a short time that it's impossible for us to measure them directly. Heisenberg's Uncertainty Principle prevents us from observing them as actual particles; and the only way we know that they are being exchanged is from the fact that the force exists. Although no physicist has found a gluon in his experiments, the power of the H-bomb is adequate testimony to the fact that the impossibly short-lived gluons which carry the strong force are very real.

Again, near Stephen Hawking's exploding black holes, we saw that electron-positron pairs can exist for the brief time allowed by the Uncertainty Principle (unless one of them falls into the hole, and the other escapes as a real particle). So the whole of empty space is filled with ephemeral particles, appearing and disappearing too fast for us to appreciate them. But they are undoubtedly there. The effect of these virtual particles on the electron moving in the hydrogen atom has been measured, and it ties in well with the theory. Wheeler and Sakharov go further, though, and say that this 'foam-like structure' of fleeting particles and antiparticles in space is what causes the force we call gravitation, rather as the structure of rubber molecules determines the springiness of a rubber band.

Black-hole doyen Hawking takes these 'vacuum fluctuations' one stage further still. The fleeting particles could briefly make up minuscule black holes – only ·000000000-0000000000000000000000001 metre across, and weighing ·0000001 gramme – which would evaporate again immediately. Although such ephemeral black holes, again so short-lived that we cannot detect them directly, may seem inconsequential, Hawking thinks that they may have profound effects in the long term.

His work on black-hole theory has led him to believe that these objects are an even more fundamental limitation to our knowledge than Heisenberg's Uncertainty Principle, which itself means that we can never calculate with absolute accuracy what the outcome of an experiment will be. Hawking's 'Principle of Ignorance' states that, in a black hole, all information is lost (except for the hole's total mass, spin and charge). And the tiny, ephemeral black holes continually appearing and disappearing throughout space may, in some sense, be robbing us of some knowledge of the Universe.

It's well-known that Einstein never really accepted 201

The largest radio telescope in the world is a fixed mesh dish, 305 m (1000 ft) in diameter, suspended in a natural hollow near Arecibo, Puerto Rico. As well as straightforward radio and radar astronomy, this dish *has been used to transmit man's first intentional interstellar message into space, with sufficient power to communicate with a similar dish anywhere in our Galaxy.*

Heisenberg's principle, the idea that pure chance plays an important role in the Universe; he insisted, 'God does not play dice.' Hawking goes further than Heisenberg and, with black holes in mind, declares that 'God not only plays dice, He sometimes throws the dice where they cannot be seen.'

The laboratory of the Universe

After our excursion along the tortuous trails which physicists are now following in their search for the basic make-up of the Universe, let us turn our eyes skywards. Astronomers have benefited profoundly from physicists' discoveries: after all, physicists had predicted neutron stars and black holes long before astronomers found them; and the theory of star evolution would be in a sorry state without the physicists' knowledge of nuclear reactions. Yet it hasn't been a one-way process: astronomers have contributed to physics, too.

For a start, we would know very little about gravitation if our science were confined to Earth. Only in a cosmic-scale laboratory do we get enough matter concentrated to allow gravitation to overwhelm the other forces. Black holes are the most recent of Nature's gravitational experiments which we have the privilege to study; but astronomical tests of gravitational theory have a long history, dating back to the time when Newton compared an apple's rate of fall with the Moon's motion around the Earth.

Neutron stars – appearing to us as pulsars – are a kind of matter too dense for us to make on Earth; and supernovae are explosions of a power we can't emulate. Both are situations where astrophysicists simply take a seat in Nature's laboratory, and watch the results of an 'experiment' beyond our own capability.

Among all the interplay of physicists and astronomers, there's one fascinating topic which only astronomers can approach. Are various fundamental quantities, like the electron's charge, and the number which relates a photon's

Only recently have geologists accepted that the Earth's continents are moving; it is now thought that India and Sri Lanka, seen here from a manned Gemini capsule, were originally joined to Africa and Antarctica. Our planet undoubtedly has further surprises in store for scientists.

energy to its wavelength (Planck's constant) changing with time? Physicists must do all their experiments now, but astronomers can reach back into the past as they look to more distant objects. And by looking at the radio 21-centimetre 'line' of hydrogen, and also the optical spectral lines of distant quasars, they find that the physical constants are not changing (or if they are, it can only be in a very special way). Astronomical observations tell us that the basic particles and forces in the Universe have not altered since the earliest times.

Turning to astronomy itself, the next couple of decades should see space probes visiting all the planets (except remote Pluto); and possibly even Halley's comet, on its return in 1986. Planetologists will be able to tie down more precisely the individual quirks of the planets; and, just as fascinating, they will begin to explore the intriguingly different moons of the outer planets. The study of the Earth itself should progress, too, for it is not many years since geologists accepted that the crust of the Earth is cracked and moving about continuously in the great plates which carry the continents. We still do not know how these huge grinding chunks of crust are propelled, with sufficient energy casually to toss up mountain ranges like the Himalayas when they collide.

The Sun's missing neutrinos are coming under attack by a new generation of neutrino detectors, more sensitive than Raymond Davis' because they'll pick up the more abundant low-energy neutrinos. Astronomers should at last have some direct evidence on the thermonuclear furnace which keeps our Sun shining. This will be complemented by studies of the minute regular pulsations of the Sun's surface, which have been a subject of dispute in recent years, but which should eventually reveal some independent evidence on the Sun's structure. Both these studies of the Sun will be tests of our current ideas of star interiors, and of the stars' evolution.

Astronomers will also keep their radio telescopes tuned for signals from other civilizations. It's a search in which we can't begin to estimate the likely success rate. Certainly, if we do pick up a message from another world, it will be the most exciting thing that astronomy has wrought. Its likely consequences are a matter not for astronomers and physicists; they are the province of psychologists and sociologists!

And in this brief overview of a few of the problems and challenges still facing astronomers, let us end up in the outer reaches of the Universe. We dwelt in Chapter 10 on the difficulties which galaxy researchers are facing, such as the problem of the 'missing mass', and the frighteningly powerful quasars which some galaxies harbour within them. And galaxy formation still remains to be sorted out.

Here the radio background radiation will help. This faint whisper from space may well hold the secret to some of the hardest questions we can ask of the Universe. Already it has demolished the Steady State theory, that beautifully elegant cosmological theory which enjoined that the Universe at all times looks the same. Improved instruments are continually being built to measure just how uniformly 'bright' (in radio terms) the sky is. Astronomers expect that there should be some slightly hotter and cooler patches; for the gas didn't stay uniform, but clumped together in the clouds which eventually collapsed as galaxies. So far no one has found them, but they'll eventually tell us something new about how the galaxies formed.

Leaving aside this possible small-scale patchiness of the background radiation, there are reasons to suspect that it may not be quite the same temperature all over the sky. Remember that our cosmology showed that space is expanding like the cake between the raisins in a rising fruit cake. The background radiation is coming from all around us; and we'd expect it all to be at the same temperature only if we are at rest relative to the space immediately around us – as if we are on a raisin which is not moving through the cake. But in the real Universe, we are not at rest. Even if our Galaxy is fixed relative to its local space, the Sun is carrying the solar system around at some 250 kilometres (155 miles) per second, and so the background should seem hotter in the direction in which the Sun is heading. Calculations showed that the variation should be only a few thousandths of a degree in the background's temperature, but in 1977 a sensitive radio receiver flown on a U-2 aircraft did pick out a variation. The sky seems to be very slightly 'warmer' in the direction of the constellation Leo.

Unfortunately, this was not the direction which astronomers had predicted. If we take the result at face value, and allow for the Sun's movement, it turns out that our Galaxy is shooting through its local space at an amazingly fast 600 kilometres (370 miles) per second. In fact, all the Local Group of galaxies would have to be moving at this high speed; and so would our neighbouring huge cluster of galaxies in the constellation Virgo. Astronomers are not happy with this result. Most previous measurements had suggested that each galaxy is pretty well 'fixed' to its local piece of space. (Although distant galaxies are racing apart from each other, this is only because the space between them is expanding.) If this first result is confirmed by later experiments, it will alter astronomers' views on how galaxies and their surrounding space are related.

Stephen Hawking has shown that the generally evenly-spread background radiation proves that the Universe as a whole is not rotating. If we took a gyroscope into space, it would always point in the same direction. Hawking has now proved that, relative to such a gyroscope, the Universe has turned by an angle less than a millionth of the Moon's apparent diameter during its lifetime of 15,000 million years. Some cosmological theories link the constant pointing of gyroscopes and our ideas of inertia – resistance to acceleration – with the distant masses in the Universe; and Hawking's result is good evidence that this may be so. Such is the power of modern astronomical observation coupled to the theories of physics.

The anthropic principle

We've concentrated in this book on the obvious frontiers of knowledge: the very big, and the very small. Scientific research isn't just concentrated at the extremes, though.

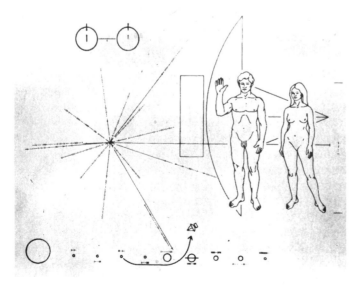

Physicists also study how atoms behave in solids and liquids, for example; and from this kind of research have come devices like transistors and integrated circuits, which are transforming our everyday lives. Chemists investigate how molecules change during the course of a chemical reaction, and also produce a wide variety of new products.

If we take an extreme 'fundamentalist' viewpoint, such research is on very intermediate ground. Only the relatively well-understood electromagnetic force is involved; the only particles are atomic nuclei and electrons, again familiar friends. What is new, what is there to find out? The answer really lies in the complexity of the interactions which occur when you put atoms together. Theoretically, it should be possible to predict the result of any chemical reaction by computing all the forces between all the nuclei and electrons involved; but in practice, this is far beyond the range of even the most modern computers. It's far more sensible to let Nature do these calculations in her own natural way: mix the contents of two test tubes, and see what results.

In the biological sciences, the complexity has gone even further beyond any hope of computing. As we saw in Chapter 5, the reactions are excessively involved, and all interrelated; biochemists are still at the stage of unravelling the secrets of life. Although experiments using Nature's materials give us the opportunity for genetic engineering, this is really just what its name implies: tinkering around with the parts which the cells have already provided. Some of the simpler of life's molecules, such as insulin, can be synthesized in the laboratory; and the future should see many more being produced – but the possibility of building a living cell from scratch is way beyond us. If we ever could, indeed, it would be a copy of what Nature has already done. To design a new plan for life, using different molecules and processes from those of Nature, is beyond our imagination.

As we saw before, living creatures are a somewhat strange development in the Universe. Starting with simple elements, and building them up by nothing more exotic than the electromagnetic force, life has evolved into systems of tremendous complexity. And not just a random complex mixture: life's processes are the pinnacle of organization. Millions of molecules in each cell are marshalled to act in unison: to contract a muscle, to pass an electrical message, to feed, to reproduce. Somehow, all the processes governed by the central DNA, and executed by proteins, have created marvellous aggregations of matter on our planet, matter which is alive: the stately grace of a Douglas fir; the gaudy beauty of a dragonfly; the thoughts and dreams of *Homo sapiens*.

It's all too marvellous, in fact: it really seems miraculous when we view life from the context of the Universe as a whole. So many 'coincidences' must have happened to allow life to exist. The forces of Nature must be balanced so

that quarks make up stable protons, and neutrons are stable in nuclei to hold the protons in against their mutual electrical repulsion. The interplay of strong and electromagnetic forces in nuclei must be just right, to ensure that the smaller nuclei, at least, are stable over thousands of millions of years. And there must be other particles which do not feel the strong force, and so can orbit the nucleus without being dragged in; these particles, the electrons, must also have an 'exclusion principle' which builds them up into different orbits – if all electrons could stay in the smallest orbit, all the elements would have identical chemistries. There had to be a particular arrangement of electrons in one element – carbon – which would let it form the long chains which are a necessity for life's complex processes.

And so the list goes on. If the weak force did not exist, the Sun could not have kept shining long enough for life to evolve. If the Big Bang itself had kept to the theorists' ideal, and contained exactly equal amounts of matter and anti-matter, there would be no matter left around after it had cooled to the annihilation point; and the Universe today would, very boringly, consist only of radiation. . . .

But let's leave it at that. If the Universe differed even slightly in the strengths of its forces, or numbers like Planck's constant of radiation, life could not have appeared, we would not be here. To explain or understand a miracle does not make it any less wonderful. Research scientists are no less 'religious' than any other group of men; when you're dealing every day with the marvels of the Universe, it's natural to feel a sense of awe.

Yet there is another way to look at the remarkable orderliness of the Universe, its curious property of allowing life and intelligence. Suppose we had an infinite number of universes to think about. If our Universe is oscillating, perpetually from 'Big Crunch-Bang' to 'Big Crunch-Bang', we could imagine these other universes to be cycles of existence before or after ours. But even if our Universe is 'open', there could be other three-dimensional universes off in a fourth dimension; we can't imagine this, but it is just as possible as stacking 'two-dimensional' sheets of paper in a third (height) dimension.

Now, each of these other universes could have different laws of physics, different fundamental particles, different forces. And the vast majority would be utter chaos. Only in a few, out of all these universes, would the initial set-up of Nature be right to allow steadily burning stars, and all the other 'coincidences' already mentioned that the presence of life requires. And so any universe where life evolved would have to be an orderly one, with a certain critical balance of forces and particles. According to this 'anthropic principle', it is no coincidence that we live in an orderly Universe.

And now we have the fascinating possibility that we can predict something about the Universe, merely from the fact that we are here. So far these predictions have been retrospective, for astronomers and physicists have pointed out that we should have been able to predict facts which are already known. American physicist Robert Dicke first showed that the Universe must be 'very large' – the distance astronomers can see must be several thousand million light years – just because we are here. For life's complex processes, we need elements heavier than hydrogen and helium from the Big Bang. Since these are made in stars, and star lifetimes are very roughly a thousand million years, the Universe must have been expanding for this time by the point at which life appears.

Stephen Hawking and his colleagues have popularized the anthropic principle in astronomical circles. They were particularly intrigued by the uniform background of radio noise. Wherever we look, the radio background is at the same temperature (ignoring our Galaxy's motion). Yet at the time the radiation left this early gas, when the age of the Universe was only around a million years, the different parts of the Universe we see in the sky were more than a million light years apart from each other. Each part of the gas could have had no knowledge of what temperature the rest of the gas had, and we should expect the sky to vary wildly in temperature from place to place.

Stephen Hawking and Barry Collins argue that only in those very few Universes where it happened that the gas was all at the same temperature could galaxies form; and without galaxies, there would be no stars, no planets and no life. 'On this view', they say, 'the fact that we observe the Universe to be isotropic would simply be a reflection of our own existence.'

The unfinished journey
The dedicated researchers who spend their working lives

A spiral galaxy, NGC 4565, seen edge-on. Astronomical observations, like those of fundamental physics, show a basic orderliness in the Universe. Complexities arise on an intermediate scale – yet complexity is essential to the existence of intelligent beings like Man. Photograph from the Mount Wilson and Palomar Observatories.

answering just a fraction of our questions 'why?' will find plenty more to do. Whatever one's research field, there's the thrill of finding out something new. Even if it's only a very minor discovery, a scientist can feel 'I am the first ever to know such-and-such about the Universe.'

Science offers as much intellectual and personal fulfilment as any other human activity. And it's a field of endeavour where every little bit helps. In music, there is little point in simply writing a good tune; you must get around to arranging it or including it in a structure of some kind. And some composers have taken the view that as soon as a piece is written down, the purely *musical* idea is lost. But in science, an idea or a crucial observation from one scientist can find its fulfilment in the work of others. Science is a far more social pursuit than composing or writing, for collaboration and the discussion of ideas are its lifeblood.

The domain of science has gradually broadened. Fifty years ago, the idea that we could look inside the proton, talk sensibly about the first one-hundredth of a second of the Universe's existence, monkey around with the hereditary material of our cells, or travel to the Moon, would have seemed preposterous. One domain where science still treads lightly, however, is that of our mind's activity. Inroads into psychology, behavioural sciences and the biochemistry of the brain are all whittling away the unknown within ourselves, as are the even more tentative investigations into parapsychology. Yet the human mind is a fearsomely complicated place. Everything which is not 'science' is here: love, joy, ecstasy, the darker emotions and the still mysterious states of trance – and possibly psychic powers? The interplay of science's deductive approach with such fundamental human traits promises to be a fascinating drama. We may be getting back to the attitude of earlier peoples, who made no distinction between science and philosophy.

So how far have we come from Empedocles, with his simple world of four 'elements' and two 'forces'? In practical terms, a very long way. Our improved understanding of matter has brought electricity in our homes, sources of power beyond Empedocles' understanding, new materials, the technology to explore space – in fact, all the advances which make our world so different from that of the ancient Greeks.

But in dissecting matter down to its finest constituents, we have left common sense far behind. Unlike Empedocles' earth, water, air and fire, we can't have an instinctive feel for quarks and leptons. They behave in seemingly bizarre ways, which physicists can only sensibly express as mathematical equations. There's a sad lack of empathy with our Universe's building bricks; even though that is hardly surprising when their realm is so different in size from ours. Nevertheless, physicists' attempts to explain both particles and forces as combinations of a fundamental 'order' and 'disorder' are strikingly reminiscent of Empedocles' 'forces' of love and hate, sometimes translated as 'strife'.

We occupy a middle ground. Each of us is an interplay of Nature's fundamental particles and forces – entities which we can't understand intuitively, but which band together in their millions of millions of millions to make the world about us and ourselves. In an almost magical way, the uncommon sense of the utterly small builds up into common sense of the everyday world.

We are a minuscule part of an infinite Universe, insignificant mites on a small planet circling an average star in a typical galaxy. Gaze out on a clear, starlit night, and feel how remote is the rest of the Universe, and how small is Man. Yet we have our intellect: with diligence and hard thought, we have plumbed the Universe, from the unimaginably small to the infinite beyond comprehension. Though science continues its quest for new answers to new questions, we are finding the framework linking the very small and the exceedingly large; the veils are beginning to fall away. We are answering the ancient questions; we are beginning to understand the Universe.

207

Glossary

absolute zero The lowest temperature theoretically possible. written 0K (zero Kelvin), it is equal to $-273 \cdot 15°C$ or $-459 \cdot 7°F$.

absorption spectrum A spectrum of lines or bands which is produced when light is passed through a gas into a spectroscope. The lines are related to the emission spectrum of the gas.

acceleration The rate of increase of velocity; that is, the increase of velocity per unit of time.

accelerator Physics: a device which increases the kinetic energy (motion) of a particle by means of an electric field.

alchemy A pre-scientific art, flourishing from about 500 AD until the late Middle Ages, which sought to transmute base metals into gold, achieve human immortality, etc. by means of astrology, mysticism, and so forth. The ancestor of modern chemistry.

amino acids A group of about twenty acids occurring in nature which link together to form chains called proteins. They are essential to life; but the *essential* amino acids are specifically those which the organism cannot synthesize and must obtain from its environment (eight in man).

angular momentum The product of moment of inertia and angular velocity. The moment of inertia of a body about an axis is the sum of the products of the mass of each element of the body and the square of its distance from the axis; angular velocity is the rate of motion through an angle about the axis. See also **spin**.

antimatter Matter composed of antiparticles.

antiparticle A particle having the same mass but exactly opposite characteristics (electric charge, etc.) as an 'ordinary' counterpart. An encounter between a particle and its antiparticle results in instant annihilation of both, releasing energy.

asteroid One of several thousand planetoids, or minor planets, the largest only 620 miles (1000 km) in diameter, which orbit the Sun between the orbits of Mars and Jupiter.

astrology The ancient art of predicting human destiny by observation of various heavenly bodies.

astronomical unit The mean distance from the centre of the Earth to the centre of the Sun ($1 \cdot 495 \times 10^{11}$ metres; about $92 \cdot 9 \times 10^{6}$ miles).

astronomy The study of the heavenly bodies, their positions relative to each other, their motion and their nature.

astrophysics The branch of astonomy concerned with the composition of the heavenly bodies, and the interaction between matter and energy within them and in the space between them.

atom The smallest part of an element.

atomic bomb A weapon working on the fission principle, which brings together two masses of a suitably fissile substance, each less than a critical mass. The combined mass is greater than the critical mass for nuclear fission, and an uncontrollable chain reaction takes place, liberating vast amounts of energy.

baryon A class of subatomic particle of relatively high weight, such as proton or neutron, composed of three quarks.

binary Dual; involving a pair. Astronomy: a system of two heavenly bodies, revolving around their common centre of gravity, hence 'double star'.

binding energy The energy which holds together an atomic nucleus. The sum of the mass of the particles in a nucleus is greater than the mass of the nucleus, for example; the difference is the binding energy, according to Einstein, who demonstrated that mass and energy were different states of the same thing. Thus binding energy was initially called the 'mass defect'. The binding energy is also the amount of energy which would be necessary to decompose the nucleus. If all the hydrogen nuclei in a glass of water could be combined to form helium nuclei, enough energy would be liberated to drive an ocean liner across the Atlantic; unfortunately, the technology for a controlled fusion reaction has yet to be developed.

biochemistry The chemistry of living matter, a specialization of the study of biology.

biology The study of life, having two main branches: *botany* and *zoology* (plants and animals), as well as several specialized related subjects.

black hole A body so condensed that its escape velocity is greater than the speed of light, so that anything falling into a black hole cannot escape, not even light. Black holes can form as the last stage of a star's evolution; much larger ones may occur at the centres of galaxies, and very small ones may have survived since the beginning of the Universe.

bond Link between atoms in a molecule, caused by one of several kinds of transfer of electrons. Also called *valency* bond. Some atoms, e.g. carbon, can cause double and triple bonds as well as single ones.

cell The smallest unit of life. Micro-organisms, such as bacteria, comprise just one cell, while a human being is a combination of almost a million million cells.

charge Electric charge is the amount of (static) electricity on a body. Ordinary matter which contains an equal number of electrons and protons is said to be electrically neutral, but that which contains an excess of electrons is said to be negatively charged (or positively charged if deficient in electrons). Although this terminology is arbitrary, electrical charge is a basic property of the fundamental particles making up matter.

chemistry The study of the composition of matter, and the effects of substances on one another.

coherent Of electromagnetic radiation: when all waves are in step with one another, as in a laser beam, but not in ordinary light.

comet A hazy gaseous cloud, surrounding a small nucleus of solid ice, which follows an elongated orbit about the Sun. Brighter comets have a tail which always points away from the Sun.

compound Of lenses: a system of more than one lens. In chemistry: a substance containing two or more elements whose atoms are linked by chemical bonds.

conservation In physics, the principle that total quantity of energy, momentum etc. of any system which is not subject to external action remains constant. For example, the law of conservation of momentum says that the momentum of two bodies before impact is equal to their total momentum after impact. The law of conservation of matter, which states that matter can never be created or destroyed, has had to be qualified since Einstein. Einstein's theories have given rise to the law of conservation of mass and energy, which says that the sum total of energy and mass times the velocity of light squared (mass-energy) is constant for any system.

constellation A group of stars whose outline describes something in the sky, e.g. Scorpius, the Scorpion. It is now known that the stars making up the outline of a figure are usually at very different distances, but the constellations are still used to describe the location of an object in the sky.

cosmic Of the Universe, as distinguished from Earth.

cosmology Study of the origin, history and nature of the Universe.

crystal Substance solidified in a definite geometric shape. Most pure substances can be obtained this way; solids which will not form crystals are called *amorphous*. The characterististic arrangement of atoms in a crystal is called a *lattice*.

current Movement of electric charge through a conductor, measured in amperes (A). Electric current is carried by the movement of the negatively-charged electrons in the conductor.

decay Of a radioactive substance, its transmutation into another substance. Also, the lessening of intensity of radioactivity as this transmutation takes place. Of subatomic particles, the transformation of unstable particles to more stable ones.

density Mass per unit volume. When measured in grammes per cubic centimetre, this equals the specific gravity of a substance.

discrete Separate, distinct, e.g. electromagnetic radiation is emitted in discrete quanta (packets), called photons.

DNA Deoxyribonucleic acid. Long threadlike molecules which store genetic information, found in the chromosomes in a cell's nucleus.

Doppler effect The apparent higher frequency of waves of light or sound emanating from an approaching object, or lower when it is receding.

dust Cosmic dust, very small solid grains found in interstellar space. Absorption of light by the grains interferes with astronomical observation of very distant objects.

electricity General term for phenomena caused by electric charge, static (see **charge**) or in motion (see **current**).

electromagnetism General term for the field of study of electrical and magnetic phenomena. An *electromagnetic* device makes use of both, such as an electric motor, in which the current flow sets up a magnetic field which causes the rotor to turn, or a generator (dynamo), in which external power turns the rotor, and the resulting magnetic field generates a current flow.

electron Fundamental particle, one of the lepton family. It orbits the nucleus in atoms, creates chemical bonds between them, and its movement is responsible for electrical phenomena.

element A substance consisting of atoms all of the same kind.

emission spectrum The spectrum found when light coming from a glowing gas is examined by a spectroscope; consists of bright lines on a black background.

energy The capacity or potential capacity for doing work. Energy can exist in any of several forms, and can be converted by appropriate means from one form to another.

energy state A configuration of an atom, nucleus, etc. in which it has a particular energy. Only certain energy states are allowed by the quantum theory; in the case of electrons in atoms, these correspond to orbits of different size.

entropy Measure of randomness, or inability to do work. The entropy of a closed system is always increasing; that is, energy always changes from a useable to a non-useable state, never the reverse.

enzyme Any of many complex protein substances made by living cells and having the effect of facilitating the chemical processes which are essential to life. As they are not changed or consumed in the chemical reactions (like any catalyst), they are effective in very small quantities.

evolution Biology: the theory that species arise by accidental genetic mutation, some of which are beneficial and allow the survival of the new, fitter species at the expense of the original. Astronomy: the changes a star, galaxy, etc. undergoes as it ages.

excite Addition of energy to a nucleus, an atom or a molecule, changing its energy state.

field Force-field; the region in space in which a force operates between two or more bodies. A body with mass has a gravitational field; it may also have an electrical and/or magnetic field; the nuclear forces create 'fields' only at subatomic distances.

fission Splitting, as in nuclear fission, where a heavy nucleus is split into two smaller ones, releasing energy (see **binding energy**).

force An outside agency capable of changing the rate of motion of a body. There are four forces in nature, according to modern physics: the electromagnetic force, gravitation, and the strong and the weak nuclear forces. These forces are thought to be different manifestations of a single basic force, and their unification is one of the goals of modern physics.

free Chemistry: an uncombined element. Physics: a *free electron* is not attached to an atom but can move in an applied electric field.

frequency Number of repetitions of a cycle in a given time. For example, the frequency of alternating current is 50 Hz (cycles per second) in the UK, 60 Hz in North America.

fusion Melting, or mingling of melted substances; hence use of term for *nuclear fusion*, the combining of atomic nuclei of lighter elements to make heavier ones. If the binding energy of the product is greater than the total binding energy of the two lighter atoms, some of the mass appears spontaneously as energy (see **mass-energy equivalence**); this is the process which goes on in stars and in the hydrogen bomb.

galaxy A huge cluster of gravitationally-bound stars, often with some interstellar gas and dust. The Sun is one of some 10,000,000,000 stars making up a typical galaxy called the Milky Way Galaxy, or simply the Galaxy.

gamma rays Electromagnetic radiation with the shortest wavelength of all, emitted by the nuclei of radioactive atoms in the process of decay. Very dangerous because of their high energy.

genetics Branch of biology dealing with heredity, the resemblances and differences between related organisms, variation and evolution.

giant A bright star of low average density, many times bigger than the Sun. A late stage in any star's evolution.

gluon A particle which transmits the strong nuclear force. Current theory postulates eight gluons, although they have not yet been detected in experiments.

gravitation The attraction of each particle of matter for every other particle. Gravitation is the weakest of the four forces known to physics, but it is extremely important in astronomy.

gravity Specifically, the gravitational force exerted by the Earth on any matter within its gravitational field.

hadron Class of subatomic particles constructed of quarks; it includes both baryons and mesons.

halflife Of a radioactive element: time taken for half its atoms to decay.

heat death The theoretical eventual state of the Universe, when entropy has run its full course: energy (or heat) will be evenly distributed so that everything will be at the same temperature. Predicted by the Second Law of Thermodynamics, if the Universe is a closed system.

holography A method of recording and reproducing a three-dimensional image (hologram) using a photographic plate and two beams of coherent light. The wave characteristics of the light are recorded, rather than its intensity, as in ordinary photography.

hydrogen bomb The thermonuclear bomb (or hydrogen bomb, or H-bomb) is a nuclear fusion device. It comprises a fission bomb surrounded by a lining of hydrogen (and often lithium) isotopes; the explosion of the fission bomb results in extremely high temperatures which cause fusion of the hydrogen nuclei to form helium nuclei, releasing even greater amounts of energy.

inertia Tendency of a body to preserve a state of rest or of uniform motion in a straight line.

infra-red Electromagnetic radiation with slightly longer wavelength than light, given off by warm objects and having a heating effect.

interference The addition or combination of waves, such as sound or electromagnetic waves. This phenomenon results in extra 'beats' in musical interference, or 'whistling' on a radio receiver in the case of interfering radio waves.

ion Electrically charged atom or group of atoms.

isotope One of several forms of atoms of an element differing in the number of neutrons in the nucleus, but chemically identical.

isotropic Having the same physical properties (shape, etc.) in all directions.

kaon See **meson**.

laser Light Amplification by Stimulated Emission of Radiation, a device producing an intense beam of coherent light by exciting atoms to emit photons of identical wavelength.

lattice Term for the regular pattern of atoms in a crystalline solid, e.g. crystal lattice.

law Of Nature: a statement which describes an inevitable sequence of specified conditions or phenomena. Scientific laws are generalizations based on many experiments; they may be disproved or extended by the results of subsequent experiments.

lepton One of the two 'families' of basic subatomic particles (the other being the quarks). The most important lepton is the electron.

life form A form of matter including complex carbon compounds and exhibiting continuous change; that which we call 'alive'.

light year The distance that light can travel in a year, used to measure the great distances outside our solar system.

magnetism An aspect of the electromagnetic force. Moving electric charges produce a magnetic force in addition to their normal electric force. In a *magnet*, the force is due to an unbalanced electron spin in the inner electron orbit of an atom in an element. Normally these are distributed at random, but if they are aligned the substance becomes *magnetic*.

magnitude System of classifying stars according to their apparent brightness.

maser Microwave Amplification by Stimulated Emission of Radiation, a method of exciting atoms to produce a coherent beam of a certain wavelength. Similar devices can be used to produce such a beam in a visible wavelength, in which case it is called a **laser**.

mass Amount of matter an object contains, hence the amount of inertia. Unlike weight, mass does not vary with the force of gravity.

mass-energy equivalence The mathematical relationship between mass and energy, if one is changed into the other. $E = mc^2$; or Energy equals mass times the speed of light squared.

matter That which has mass; in other words, that which is not pure energy (like radiation).

meson Class of subatomic particle consisting of a close quark-antiquark pair. Mesons include the K-meson (kaon), pi-meson (pion) etc; but mu-meson (muon) has been reclassified as a lepton.

microwaves Electromagnetic radiation with wavelength between radio waves and the infra-red.

molecule A combination of atoms joined by chemical bonds; the smallest part of a substance (an element or a compound) retaining the chemical properties of that substance.

momentum Mass multiplied by velocity.

muon An unstable fundamental particle, one of the leptons, 207 times heavier than the electron.

nebula A cloud of dust and gas in which stars are forming.

neutrino Fundamental particle (one of the leptons) with no mass and no charge; very difficult to detect. Two or three kindsof neutrinos exist, associated with the electron, muon and possibly with the recently discovered tau-lepton.

neutron Subatomic particle (classified baryon) with no charge, consisting of two 'down' and one 'up' quark. Neutrons are found in the nuclei of all atoms except hydrogen.

neutron star A small, dense body formed at the end of a massive star's evolution when it has consumed all its nuclear fuel and collapses due to gravitation, becoming so dense that it contains only neutrons. Rotating neutron stars emit radio pulses, and are known as pulsars.

nova A star ejecting part of its material in the form of a gas cloud, becoming more luminous during the outbursts. Nova outbursts are thought to originate from a white dwarf star in a close double star system.

nuclear Having to do with the atomic nucleus; e.g. reaction, explosion, fission, fusion, etc.

nucleus Physics: the central mass of an atom, containing nearly all of its mass, consisting of protons and neutrons (except for hydrogen) and carrying a positive charge. Biology: the central part of a cell, containing the hereditary material DNA. Astronomy: the centre of a galaxy.

organic Of chemical compounds: containing carbon, and usually connected in some way with life processes. Some very simple compounds such as metal carbonates are arbitrarily excluded.

parity A law of conservation, also called space-reflection symmetry, which says that there can be no distinction between left and right as far as the laws of physics are concerned. This law is violated by particles reacting by the weak nuclear force. This was proven by Chien-Shiung Wu in 1956. She froze cobalt 60 atoms to within a hundredth of a degree of absolute zero, and placed them in a magnetic field; their nuclei threw electrons mostly upwards, (with left-handed spin) rather than up and down at random.

parsec A unit of measure of astronomical distance equal to 3·26 light years. Based on the parallax phenomenon, which is the apparent displacement of an object caused by the change of point of observation.

particle Term generally used to mean a particle of subatomic size, including any of the particles of which atoms are made (proton, neutron, electron, etc.), various others usually observed separately (photon, neutrino, etc.) and the quarks, of which protons, neutrons, etc. are made but which have yet to be observed in experiments. All subatomic or 'fundamental' particles are either force particles (e.g. the photon), leptons or are made of quarks.

phase State of wave motion: two points are in phase if their cycles are taking place in the same direction at the same time.

photon A particle of electromagnetic energy, having zero mass and travelling at the speed of light. In various facets of its behaviour it must be considered as a wave and as a subatomic particle at the same time. It also conveys the electromagnetic force, as a *virtual* particle.

physics The study of the various states of matter, including energy.

pion See meson.

plasma Ionized gas in which the proportion of positive to negative ions is more or less equal. Also the very hot ionized gas (10,000,000°C) in which nuclear fusion can take place. Plasma is electrically neutral and highly conductive.

plate tectonics The theory of continental movement, which considers the continents as 'rafts' travelling on larger plates sliding over the Earth.

polarization Of light: vibration of the transverse waves of light in a particular plane, instead of at any angle; some materials can be used as polarizing filters to pass light with waves in a certain plane and exclude other light.

positron Antiparticle corresponding to the electron; it is the same mass as an electron, but is positively charged.

principle A general scientific rule governing fundamentals rather than details, as opposed to a law of nature, e.g. Heisenberg's Uncertainty Principle compared to the law of conservation of mass and energy.

prominence A cloud of luminous gas projecting above the Sun's surface.

protein Complex organic compounds whose molecules are built up from smaller molecules of amino acids; essential for all living matter. Plants can synthesize them; animals must get them, directly or indirectly, from plants.

protogalaxy An enormous cloud of dust and gas in the process of forming a galaxy.

proton Subatomic particle composed of three quarks (two 'up' and one 'down'). Protons are found in varying numbers in all atomic nuclei; positively charged.

pulsar Rotating neutron star, emitting radio, light or X-rays. The emission appears to 'pulse' as the star turns.

quantum, plural quanta Individual, smallest possible unit of energy. Quanta of electromagnetic energy are also called photons. Quantum theory is the original form of the theory that energy exists in quanta, rather than continuously, put forward early in this century by Max Planck (1858–1947).

quark One of the two 'families' of fundamental particles, the other being the leptons. A mathematical scheme called *unitary symmetry*, a method of classifying properties of behaviour, was originally used by Murray Gell-Mann to postulate that only three particles, arranged in different ways, could be the constituents of the fundamental particles known as baryons (proton, neutron, etc.) and mesons. But there are now four quarks known ('up', 'down', 'strange' and 'charmed'); another ('bottom' or 'beauty') has turned up in recent experiments, and a sixth ('top' or 'truth') is suspected. No quarks have been successfully isolated as yet.

quasar Very distant, highly luminous astronomical object. Quasars are almost certainly the very active centres of remote galaxies; the energy source is unknown, but may be gas falling into a very large black hole.

radiation Emission of waves or particles at a regular pulse or wavelength. Electromagnetic radiation is a form of energy which includes many phenomena from radio waves to gamma rays (from radioactive substances), including light; it is generated, detected and classified according to its wavelength. Gravitational radiation should also exist, and several experiments are searching for it at present.

radioactive Adjective describing the behaviour of certain heavy elements, whose nuclei are decaying and emitting particles and radiation. In concentrated form this radiation is poisonous because of its extremely high energy.

radioastronomy Use of various types of receivers (instead of optical telescopes) to gather information from outer space in the form of long-wavelength electromagnetic radiation (radio waves). The energy received is extremely low, and the science is only about fifty years old, but the information is of incalculable value.

radioisotope Radioactive isotope of any element. May be natural or artificially produced. Used as source of radiation for medical or industrial purposes, etc.

radiotelescope Very sensitive aerial (antenna), used in radioastronomy. Often in the shape of a large dish.

rarefy, rarefaction To reduce pressure, cause a substance to become less dense. Opposite of compression.

ray Strictly speaking, the path along which a radiation travels; also loosely used to describe the radiation itself.

reaction Chemistry: interaction of two or more substances, causing changes in them.

red shift Increase in wavelength of light (so that it becomes more red) of stars and galaxies moving away from the Earth at high velocities. Also called Doppler shift.

refraction The diversion of the path of light (or other radiation) when it passes from one medium to another.

relativity Two theories put forth by Einstein in 1905 and 1916, which have involved modification of the laws of physics. Based on the fact that the speed of light is always the same regardless of the relative motion of the observer, these theories describe how length, time and mass are affected by motion, and include a new theory of gravitation.

resolution Degree of power of a lens or set of lenses to distinguish between two very small objects.

RNA Ribonucleic acid. Substance in living cell which picks up and transfers hereditary information stored in DNA.

shell, electron Term used in quantum mechanics for the series of concentric circles described by electrons orbiting a nucleus at various distances.

singularity The theoretical centre of a black hole, in which matter is compressed until it occupies zero space. Even more esoteric theory postulates a singularity without a black hole around it, in which case it is called a naked singularity.

spectroscopy Science of analysis of the electromagnetic spectrum of an object.

spectrum A resolution or separation of electromagnetic waves according to wavelength. The best known example is the selection of colours seen in a rainbow, and when white light is passed through a prism. The entire electromagnetic spectrum extends from long radio waves to short-wavelength gamma rays.

spin The angular momentum of subatomic particles which revolve about an axis within themselves is called spin. It comes in discrete units; the quarks and leptons have half a unit and the force particles one or two units.

storage ring Device used in particle physics to keep particles circulating while beams are built up to high intensities.

strong nuclear force The strongest of the four forces in nature, responsible for controlling the interactions of nuclear particles. 100 times more powerful than the electromagnetic force.

sunspots Large, dark patches on the surface of the Sun, which appear to cross its surface as it rotates and which occur in their greatest numbers about every eleven years.

supernova The explosion of a star, a rare event when the star becomes a thousand million times brighter than the Sun. When a massive star runs out of its nuclear 'fuel', it can collapse on itself because of its own gravitation with such force that high temperatures and runaway thermonuclear reactions follow.

theory An idea or system of ideas, based on general principles but explaining specific phenomena. If new phenomena appear which do not fit the theory, the theory must be modified or discarded.

thermodynamics Literally means 'heat movement'. The science and study of heat, its transfer, behaviour, etc. Founded by French physicist Sadi Carnot (1796–1832).

thermonuclear Literally means 'heat of the nucleus' and is used to describe nuclear fusion, as opposed to fission. A very high temperature is necessary for fusion.

transmutation Changing of one element into another. Once the goal of alchemy, it was next assumed to be impossible. Radioactive elements are doing it naturally, though slowly. The basis of the nuclear power industry and of nuclear weapons.

transparent Physics: allowing electromagnetic waves to pass through. For example, a thin layer of concrete is transparent to radio waves but not light.

ultra-violet Electromagnetic energy with higher energy (shorter wavelength) than light but below X-rays in the spectrum.

uncertainty principle First postulated by Werner Heisenberg, who said that due to the dual (particle-wave) nature of matter, both position and momentum of a particle cannot be known exactly; the more certain is the position the less certain is the momentum. Similarly, the energy of particle cannot be known at a precise moment in time.

vacuum The complete absence of matter. No perfect vacuum is known to exist; even interstellar space is not a complete vacuum, but contains very tenuous hydrogen gas.

valency The power of an atom to combine with other atoms. See also **bond**.

velocity Speed in a particular direction, measured as length per unit of time, as 'miles per hour'.

virtual (particle) A fundamental particle conveying a force. The electromagnetic force for example, as opposed to radiation, is carried by photons which are 'virtual' because they exist for too short a time to be detected.

wave A periodic disturbance carrying energy. It can be in a medium, such as water or air, or it may be composed of energy itself, such as electromagnetic energy, in which case it can cross a vacuum. The period (distance between waves) is called the wavelength.

wavelength The distance between successive 'crests' of a wave.

weak nuclear force The weakest of nature's four forces except for gravitation. Responsible for the behaviour of decaying particles. A theoretical relationship has been found between the weak nuclear force and the electromagnetic force; this is the first step toward a unified theory of all the forces.

weight The force of gravitational attraction on a mass, especially when it is on Earth. The weight of a body would change, but not its mass, on another planet.

white White light: a light which can be resolved into all the colours of the spectrum. White hole: theoretical opposite of a black hole, a fountain of energy, now believed not to exist. White dwarf: a small, dense star of low luminosity. Because of its small size it is very hot and so appears white.

X-ray A part of the electromagnetic spectrum having higher energy (shorter wavelength) than light and ultra-violet, but lower than gamma rays. Has an effect similar to that of light on photographic paper, but passes through matter to an extent depending on the density and the atomic weight of the matter. Hence it is possible, for example, to take X-ray pictures of the bones of a living creature. Prolonged exposure to X-rays can be dangerous because of their high energy.

Index

Figures in **bold type** indicate entries in the glossary. Those in *italics* indicate illustrations. 'a' and 'b' mean left and right halves of pages respectively; '(g)' means glossary; 'U' means the Universe.

BIBLIOGRAPHY

Many research papers and specialist books and reviews have been consulted in preparing this book. Among the publications which have been particularly useful are:

Annual Reviews of Astronomy and Astrophysics
Astrophysical Journal
Physics Bulletin
Quarterly Journal of the Royal Astronomical Society
Sky and Telescope
Gravitation, by C. W. Misner, K. S. Thorne and J. A. Wheeler (Freeman 1973)

Readers who want to keep abreast of the latest developments can find up-to-the-minute reviews in *New Scientist* (UK) and *Scientific American*. These generally require a little scientific background. Further reading at roughly the same level as in this book, but dealing in more detail with specific topics, can be found in the following books:

General astronomy
  Constellations, by J. Klepestra and A. Rükl (Hamlyn 1977)
  Norton's Star Atlas, by A. P. Norton (Gall and Inglis 1973)
  Cambridge Encyclopedia of Astronomy, edited by S. A. Mitton (Jonathan Cape 1977)
Fundamental particles and forces
  Key to the Universe, by N. Calder (BBC 1977)
Biography
  Biographical Encyclopedia of Science and Technology, by I. Azimov (Doubleday 1964; Pan (UK) 1975)
The Earth
  Restless Earth, by N. Calder (BBC 1972; Futura 1975)
Stars and galaxies
  Violent Universe, by N. Calder (BBC 1969; Futura 1975)
Cosmology
  The First Three Minutes, by S. Weinberg (Deutsch 1977)
Others:
  Radio Astronomy, by F. G. Smith (Penguin 1975)
  Matter, by R. E. Lapp (Time-Life Books 1969)
  The Cell, by J. Pfeiffer (Time-Life Books 1969)
  The Cosmic Connection, by C. Sagan (Hodder and Stoughton 1974)
  Signs of Life, by I. Ridpath (Kestrel 1978)